Mariana Lucas Moreira
Marcos Anciuti
Fabiana Andersson

An alternative and dynamic way of teaching oligochaetes

Mariana Lucas Moreira
Marcos Anciuti
Fabiana Andersson

An alternative and dynamic way of teaching oligochaetes

Science is Life

ScienciaScripts

Cover image: www.ingimage.com

This book is a translation from the original published under ISBN 978-3-330-75988-6.

Publisher:
Sciencia Scripts
is a trademark of
Dodo Books Indian Ocean Ltd. and OmniScriptum S.R.L publishing group

120 High Road, East Finchley, London, N2 9ED, United Kingdom
Str. Armeneasca 28/1, office 1, Chisinau MD-2012, Republic of Moldova, Europe
Printed at: see last page
ISBN: 978-620-8-31407-1

SUMMARY

Presentation 2
Chapter 1 INTRODUCTION 4
Chapter 2 Science Teaching and the Study of Oligochaetes.... 24
Chapter 3 Science Teaching: Theory vs. Practice 46
Chapter 4 Where We Started From, What We Uncovered 71
References.... 77

Learning is the only thing that the mind never tires of, is never afraid of and never regrets."

Leonardo da Vinci

PRESENTATION

The purpose of this work is to provide students with a way of learning about the biological role and reproduction of oligochaetes, based on practical tests. As well as diagnosing the didactic methodologies applied by science teachers in the municipality of Piratini/RS, this study linked theory with practice through the practical activity of vermicomposting with fruit waste and dairy cattle manure. The target audience for this research was 24 teachers from the municipal and state school systems in the municipality of Piratini/RS, and 42 students from the 6th to 9th grades of primary and agricultural education at the Alaôr Tarouco Municipal Agricultural Primary School. Prior to the practical activity, the students and teachers answered a questionnaire with open and closed questions about the biological role and reproduction of oligochaetes, as well as their opinions on practical activities. After applying the questionnaires, the practical activity was carried out, understood here as a didactic alternative for studying oligochaetes. After the practical activity, the students wrote a report on the activity, which we compared to the answers they had given in the questionnaires. It's important to note that this practical activity has led to the production of a documentary, which will be made available in digital format and on the web, and which includes testimonies from students and teachers. Although teachers consider practical activities to be essential to students' education, they still use theoretical tools in the classroom, such as books and, in some cases, videos. From the analysis of the information collected, this results in a low level of appropriation of the content by the students. However, when the content seen in class was continued with the practical activity, it was possible to observe a better assimilation of the knowledge by the students, given that they reported in detail on the operations they had worked on. It is therefore understood that the proximity of practical activities to students' daily lives broadens their worldviews, encouraging them to seek solutions to possible problems in our societies. For this reason, it is believed that the expansion of practical classes provides a better understanding of the content seen by the students in theoretical classes, especially those related to their personal formation.

CHAPTER 1 INTRODUCTION

This section outlines the ideas that led the student who is speaking here to the Master's Degree in Science and Technology in Education at the Federal Institute of Rio Grande do Sul - Pelotas Visconde da Graça Campus. For this reason, the following proposal is to bring out the main motivations for carrying out and conducting this research, which is based on the teaching and learning of "Oligoquetas", permeated above all by theoretical and practical activities. Next, the research objectives and hypotheses will be described so that we can present the trajectories of our immersion in the field. Finally, we will briefly describe the chapters that make up this dissertation.

1.1 Context

The task of contextualizing, in my opinion, is not worthy of simplification. However, despite highlighting here some of the paths I have traveled over the years on the path of teaching and education, I want to give the reader a glimpse, albeit imperfectly - since this material focuses above all on writing - of some of the places I have traveled, the choices I have made, the obstacles I have overcome and those that, due to the size of the moment, I have turned away from. And all these choices, to a large extent, have their roots in long days of reading, in the uninterrupted search for knowledge.

So, notwithstanding the importance attributed to the empirical knowledge of this author, it is also important to highlight the readings, books and materials that led me to critical and logical thinking, until this period in which not only ideas are built, but especially considerations about theoretical and practical teaching, in this case in particular, of Oligoquetas. In the meantime, and as a first step, the following proposal is to present the facts and situations that motivated me to begin my master's studies on the Master's Course in Educational Sciences and Technologies at the Federal Institute of Rio Grande do Sul - Pelotas Visconde da Graça Campus. Then some of the references that guided the beginning of this study will come into play.

1.1.1 Presentation

It would be imprudent of me to assume that research is something genuine in my being. I say imprudent because research itself is perceived primarily as an object of analysis. Even so, I became interested in it at the beginning of my school experience, because when I was doing individual or group work, I always looked forward to success, in order to receive due credit for the depth of the presentations[1] . Time passed, but the research remained, broadening my horizons and leading me closer to it.

In 2006, the year I finished high school, after a long period of construction and deconstruction, the Catholic University of Pelotas (UCPEL) finally brought the Biological Sciences degree course to the city of Piratini/RS - my hometown - as an extension program. This made it possible not only to continue my studies, but also to immerse myself in the teachings involved in scientific research.

From the very first classes, I can say that there was a great synchronicity between what I wanted to do professionally and the subjects offered in the course, especially those that referred to the environment. So much so that whenever any opportunity arose to debate, present proposals, including intervention, I did my best to take part.

For these reasons and others, I readily accepted an invitation from a fellow student, who works for a water treatment company, to research the daily lives of some families and their relationship with water. From this partnership, work to raise awareness of the responsible use and/or consumption of water was developed and disseminated through lectures and/or mini-courses in schools. Little by little, our activities moved out of the classrooms and into the old streets of Piratini, which at the time had twenty thousand inhabitants (IBGE, 2010) .[2]

---

1 As far as my work and presentations are concerned, I'm referring to those I carried out during my initial training, in the first grades of elementary school, at the Ponche Verde State Institute of Education, in the municipality of Piratini/RS.

2 Currently, only nineteen thousand people live in the municipality of Piratini/RS. The reduction in its population, as is the case in almost all cities in the southern zone, except cities seen as metropolises, has its origins in unemployment, in the lack of opportunities for young people who leave in search of

Although I was a student, I soon started working as an intern at the Municipal Education Department in the municipality of Piratini. But as soon as I graduated in Biological Sciences, I started teaching in the classrooms of one of the rural schools, which mainly focus on primary education. After some time teaching in this school, I got to know the reality of schools in the "urban area". This brought new vigor to my professional practice, given the work I had done at the school responsible for my initial training. There, I had the opportunity to teach not only elementary school, but also secondary school and Youth and Adult Education (EJA).

From that moment on, I realized the burdens and bonuses of the profession. I had to prepare lessons on a daily basis for all levels, having to put into practice the concepts I had internalized during my training and permeated by my experiences and practices in "Science" and "Biology". The students of the time developed projects, created experiments, analyzed data, wrote, reported, among other activities.

However, as teaching was largely based on the traditional model, with its emphasis on content, it needed some renewal. In order to combine practice with theory, allowing knowledge to be built up with the student(s), I was then guided by the pedagogical relationship of transmission and critical assimilation. It was with this expectation that I took over the coordination of the Piratini Educational Center for Distance Learning (EAD), with two courses: Agroindustry and Biofuel, linked to IFSUL/CAVG.

It's worth pointing out that, in this case, I took part in the whole process, from the creation of the center and the courses, the selection process, enrollment and everything else that goes into the role of "coordinator". For this workload, I felt the need to create and build an environment of motivation and participation that really involved the students. This makes sense when we look at our current educational situation. We are at a time when we need to review values,

---

better life chances.

because what used to be customary has become inappropriate.

To a large extent, the expository teaching model used in most schools is not attractive to students. That said, it is urgent for teachers to create alternatives that arouse students' interest and curiosity, making them the "protagonists" of and in the teaching-learning process.

From this perspective, science teaching should be based on practical activities aimed at theoretical understanding, seeking, whenever possible, to rescue the preservation of the environment for the future. Furthermore, the school must be seen as a partner in order to make it possible to change paradigms in favor of building a common future, looking at what we have been in order to take care of what we will be.

In this regard, I'd like to mention that in all the schools where I've taught, I've done a lot of work that has made an effective impact not only on the students' school life, but also on mine. Subjects from the students' daily lives, such as "garbage", were often worked on in the classroom and, when presented at science fairs, won awards .[3]

As a result, I noticed greater student participation in the subject, with repercussions even in the school community and the general public. It's important to consider that this way of working took us out of the usual, out of books and notebooks, in other words, out of the "comfort zone", to get involved in teaching processes based on the use of other materials and the integration of subjects - my focus as a teacher.

Despite my attempts to broaden my teaching experience, science lessons are basically based on theory. This is why I am interested in creating a new didactic strategy, linking theory and practice[4] , to work on the content relating, for

---

3 As an example, I cite an essay from the "Agrinho Project" in which a certain student received a certificate and had her writing published.

4 And here we corroborate Antunes (2013), by understanding that only with diversified classes and practical activities can we build a teaching process capable of responding to the challenges and demands imposed by reality and linked to the student's daily school life, and that, in this situation, he is the essential part of the teaching-learning process.

example, to the reproduction of oligochaetes (earthworms) in fruit waste and cattle manure, by means of the Vermicomposting process[5] - a fact which, due to the ease of observing and developing the activity with students and teachers, will be better discussed below.

1.1.2 Designing the Research Problem

When we look at the title of our work, "An alternative and dynamic way of teaching oligochaetes", we can see the presence of different issues. In other words, although our research theme focuses on processes related to education, especially with regard to the integration of theory and practice, it expands to include, or seek strategies for, the everyday problems of a growing population.

Nowadays in Brazil, collected household waste contains more than 50% by weight of organic matter that has been treated improperly or disposed of incorrectly on the ground, generating liquid (leachate) and gaseous (biogas) effluents that pollute both water bodies and the atmosphere. The collection of this organic waste is often expensive and the lack of options for its disposal results in the improper accumulation of these materials in landfills, leading to the pollution of groundwater, generating various pathologies in residents near these regions (MOLINARI, 2014).

According to Scarlato and Pontin (1993), landfills favor the contamination of nearby water basins. This is mainly due to the leachate resulting from the decomposition of organic matter thrown away and which, for lack of better options, accumulates in often precarious landfills. In addition, there are certain people in these "landfills" - notoriously marginalized by society - who reuse the food found there and possible materials that can be recycled and sold. The situation these people find themselves in, as we have said, goes beyond

5 In such a consumerist reality, garbage and leftovers (food waste) become a chaos in everyday life. That's why the theme Oligoquetas (earthworms) was chosen, because their way of reproducing comes from using leftover fruit. In this way, waste can be reused, for example as fertilizer for vegetable gardens.

marginalization, as they are "easy" targets for poisoning and infection. Furthermore, places like landfills, when poorly managed, serve as a breeding ground for disease.

In our view, one of the ways to minimize the harmful effects of the incorrect disposal of organic waste is to raise public awareness about the reuse of this waste. In other words, transforming organic waste, for example, into beneficial fertilizer for growing plants. One of the ways in which this can be done is through the construction of specific spaces for the development of what we know as "vermicomposting". Vermicomposting is, in a nutshell,

> the process of recycling organic waste by raising earthworms in vermicomposting ovens, offers an important alternative to solve the problems of organic waste, such as household waste, both economically and environmentally. The end product of vermicomposting is an excellent organic fertilizer (humus), capable of improving the soil's chemical (supply, better retention and cycling of nutrients), physical (improved structure and formation of aggregates) and biological (increased diversity of organisms beneficial to the soil) attributes (EMBRAPA, 2016).

As we can see from the excerpt above, vermicomposting is a nutrient recycling system in which organic matter is digested by earthworms and excreted in the form of humus[6] . The significance of this system comes largely from the fact that a single earthworm can eat the equivalent of its own weight, reducing domestic organic "waste" by up to 2/3 (two thirds) in one (1) day. Still on the subject of vermicomposting, we can define it as an ecologically correct, economically viable and socially just procedure, since it significantly reduces the disposal of organic waste in the environment, has a low cost, does not require trained people to maintain it and can be implemented in houses, apartments and plots of land, since it doesn't need much space.

In view of the importance of vermicomposting for the natural and human

6 Humus is the result of the digestion of organic matter by earthworms. This digestion takes place through complex chemical and physical processes. Unlike conventional chemical fertilizers, humus does not pose a risk of overdose, as the plants will only absorb what they need. As a natural product, it doesn't harm the environment and helps to keep the soil's microfauna in balance. Another benefit of its use is that it prevents the compaction of clay soils and promotes the aggregation of sandy soils, as well as retaining nutrients and water more efficiently.

environment, we decided to carry out this work. With this project, we intend to present a practical and economical option for the school community and the surrounding area, in order to reduce the amount of organic waste generated not only in homes, but in the school itself, by reusing the fruit waste used in school meals. To this end, we propose to research the application of earthworms in the decomposition of the "garbage" generated at the Alaôr Tarouco Municipal Agricultural School, with the aim of reducing the disposal of these substances as a result.

But how do we get the information about vermicomposting and the need to reduce the production of organic waste across to students? How can they transform this information into new dynamics? And how can they improve not only their teaching processes, but also their experiences of preserving the environment? It is precisely the answers to these questions that we propose to discuss in this text.

As we can see in teaching practice, science teaching is basically based on discussing the content taught in theory classes, leaving the interrelationship between theory and practice on the sidelines of the teaching and learning process. Faced with this, not just us, but groups of scholars and researchers such as Dewey (1916), Freinet (1975), Carvalho (2004) and Antunes (2013) have been reflecting on the causes and consequences of these issues, seeking solutions based on an education geared towards the full participation of students.

In this context, these authors see practical activities emerging as a positive didactic strategy for the teaching and learning process, which teachers and students point to as one of the ways of minimizing the difficulties of learning theory. And we agree with them, so much so that we are convinced that renewal in science teaching is imminent.

It is of the utmost importance to develop teaching through practical lessons, as these are based, in a way, on the student's interest and reality. In our opinion, this makes the teaching and learning process more dynamic, attractive and meaningful, arousing the student's curiosity. Considering the fundamental role of practical activity for learning, the search for alternative didactic strategies is fundamental, since it is through them that physical-chemical and biological phenomena can be explored in an integral way, both from the point of view of student involvement and the conception of the practical activity itself.

On the basis of the above discussions, it is worth clarifying that this master's work is part of the Science research line, based on "Science Teaching". On the one hand, our study proposal focuses on the biological role of oligochaetes and their reproduction in the transformation of fruit waste and dairy cattle manure through the vermicomposting process. On the other hand, the appropriation of this process by the students of the Alaôr Tarouco Municipal Agricultural Primary School, based on teaching theories and practices.

Alongside this, the question of "***how can we provide students with a way of learning about the biological role and reproduction of oligochaetes, based on practical tests?***" was ***raised.*** *The* answers to this question, which is at the heart of our research problem, were largely answered in the study in question. Even so, it should be borne in mind that the discussions that will follow on from this work are based on the premise of combining theory with practice, in order to make the students' understanding of theoretical knowledge about the biological importance and reproduction of oligochaetes effective from the moment they observe it.

1.2 Objectives

Taking as its empirical universe the teachers and students of the Alaôr Tarouco Municipal Agricultural Elementary School, located in the municipality of Piratini/RS, this research chose as its theme to highlight the possible gaps between the theoretical and practical knowledge developed in science classes at this school. To this end, we set ourselves some objectives which will be presented below.

1.2.1 General Objective

- Propose teaching strategies that provide students with an effective way of learning about, in this specific case, the biological role and reproduction of oligochaetes.

1.2.2 Specific Objectives

a) To diagnose which didactic methodologies are applied in the teaching of content related to the biological role and reproduction of oligochaetes by science teachers in the municipality of Piratini/RS;

b) To assess the pre- and post-conceptions (before and after the activities related to the dynamics of vermicomposting) of the students of the Alaôr Tarouco Municipal Agricultural Elementary School about the biological role and reproduction of oligochaetes;

c) Propose a new teaching strategy to develop the content on the biological role and reproduction of oligochaetes in the transformation of waste;

d) Prepare a document based on the didactic strategy developed, to be distributed to teachers, initially in the municipality of Piratini/RS.

1.3 Hypotheses

Our hypotheses actually guide our thinking. To this end, it is important to emphasize that they were not drawn up within the rigidity of hypothetical-deductive schemes, but were conceived as guiding the approach to the subject, as we have already mentioned, of seeking strategies that provide the student with a learning path about the biological role and reproduction of oligochaetes, based on practical tests. In this sense, our hypotheses are:

a) Science teaching is based on mainly theoretical teaching schemes, which disregard or take little account of the issues that guide and involve students. In other words, rigidly-established methodologies are employed which, for this very reason, precariously encompass the situations posed in the daily lives of those (the students) who seek knowledge/training in teaching environments.

b) Although science teaching is largely based on theoretical lessons, there are a wide variety of teachers and researchers who are looking for alternatives to the rigidity of this approach. One of the proposals we've seen is for practical lessons to be held after theoretical teaching. As will be shown in this outline, we believe that the teacher's initiative to take students into everyday situations of their own experience will provide greater and better quality in the appropriation of the knowledge acquired during their formative years as students.

1.4 Work Development Methodology

In order to give an account of the research itinerary, we must first point out that we didn't use a single line of reasoning. In fact, we used a number of methodologies and tools to carry out our work. From this point of view, it is imperative to stress that we will not only talk about our conduct during the research period here, but we will also follow up on the central idea of this section, which is to guide readers along our path of elaboration and conduction of the research.

So, in a first discussion, we will focus on the delimitation of our empirical universe. Next, we will try to define the methodological scope we used in the investigation.

1.4.1 The empirical universe of the research

The work in question covers, to a large extent, the elementary school network in the municipality of Piratini/RS. This choice is mainly due to my commitment to giving back to my hometown for the knowledge I have acquired during my professional training. Within this scope, it is essential to begin the arguments in this subsection by characterizing the municipality in question, namely Piratini/RS. The IBGE Cidades website (2016) states that its settlement began

on July 6, 1789, with 48 Azorean couples on the condition that they would live and work there. When these settlers arrived, they settled in the place called "Capâo do Piratini" and founded a chapel in honor of Our Lady of the Conception, since then the patron saint of the municipality. On the occasion of the Revolution

Farroupilha, notable events took place in the municipality. On October 8, 1835, the Farroupilhas occupied the town and formed the Independent State of Rio Grande do Sul. On November 6, 1836, they formed the first government of the republic, with President-General Bento Gonçalves da Silva as its head. The seat of the State Government of Rio Grande do Sul was then called Palâcio Piratini. On the political scene, the municipality became famous during the Farroupilha Revolution for becoming the Capital of the Rio-Grandense Republic (IBGE CITIES, 2016).

Currently, the municipality of Piratini has approximately 20,000 inhabitants spread over its 3,539.688 km^2 of land. In a way, this municipality can be attributed a history marked by rurality (Fig. 01). In Sell's (2011) view, this municipality portrays, on the one hand, farming styles geared towards large tracts of land - given its incompatibility with the strategies of the period known as the Modernization of Agriculture[7] . On the other hand, with the arrival of new people from the 1990s onwards, farming styles geared towards family-based production models have gained in importance - with the massive presence of settlements created on the basis of land reform guidelines .[8]

It is interesting to note that, with the arrival of these people, not only the agrarian issue changed, but also others, such as the education sector (RIBEIRO, 2001). According to the Municipal Education Plan, there are nine elementary schools in the municipality of Piratini, two of which are in the urban area and the others in the rural area. Among the rural elementary schools, there is one that also offers agricultural education, the Alaôr Tarouco Municipal Agricultural Elementary School (Fig. 01) - the focus of our investigation .[9]

---

7 This period brought profound changes in agriculture. What had previously been practiced on large tracts of land, respecting the production cycles and specific characteristics of each plant species, as well as geographical location, were gradually replaced by processes seen as more modern, based on agricultural mechanization, the use of improved seeds, the addition of chemical fertilizers (to improve soil fertility), among others (GRAZIANO DA SILVA, 1982). However, as the municipality of Piratini is characterized by the presence of shallow soils and, in some places, high slopes, it was neither possible nor feasible to implement mechanization, for example, a fact that contributed to making the processes of modernization in the region's agriculture unfeasible (SELL, 2011).

8 According to Sell (2011), "there are a total of 15 settlements set up in Piratini, 10 of them created by INCRA, 4 created by the state and one in a partnership between INCRA and the state, making up a total of 15,256ha, which corresponds to 10.28% of the municipality's area".

9 The Alaôr Tarouco Municipal Agricultural Elementary School was founded on March 25, 1991. Its area for teaching and educational activities totals 53 hectares, distributed over a wide range of

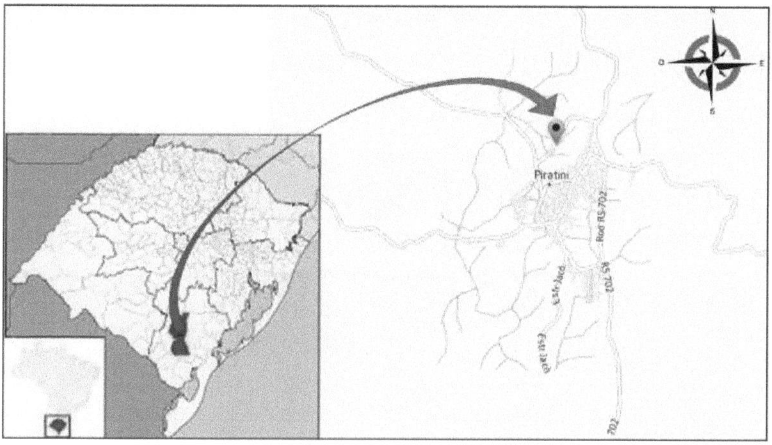

Figure 01 - Illustrative map of the municipality of Piratini with the location of the Escola Municipal Agropecuària de Ensino Fundamental Alaôr Tarouco **Source**: adapted from Wikipedia (2016) and Apontador (2016).

With regard to the choice of the Alaôr Tarouco Municipal Agricultural Primary School for the current investigation, which is to analyze the processes involved in teaching and education, through the perspectives of students and teachers, with regard to bringing theoretical lessons closer to practical dynamics, it is worth clarifying that this is mainly due to the significant range of teachers and students present there[10] . Although this school is located in the rural area of the municipality of Piratini, many of the students and teachers come from the urban area.

Also, because the school shares its land with the Municipal Garden (Fig. 02). The presence of the garden within the school's boundaries significantly provides practical experiences that go beyond the school routine for students

activities suitable for teaching and student practice. Today, the school teaches more than 100 (one hundred) students, aged between 11 and 16, enrolled in the 6th (sixth) to 9th (ninth) grades of elementary school and agriculture, the latter in the afternoon.

10 Several teachers took part in our research. Among these, it is important to note that not only those from the sciences gave us the opportunity to meet them, but also teachers from the language and exact sciences, among others. Similarly, as far as the students are concerned, it is important to bear in mind that they have empirical knowledge that corroborates their own environments, making them unique in their worldviews. This fact, in a way, clarifies the range of teachers and students we contacted who took part in the process of writing this dissertation.

and teachers, such as the production of seedlings for the municipality's plant composition, the breeding of oligoquetes to make organic fertilizers, work on behalf of the environment, among others.

Figure 02 - Area of operation of the Municipal Garden of the Municipality of Piratini/RS, with emphasis on the seedling production beds.

Source: author's archives (2015).

In view of the above, our empirical universe of research is, to a greater extent, the students and teachers of the Alaôr Tarouco Municipal Agricultural Primary School, the agricultural technician responsible for the Municipal Garden and also some teachers from the municipal and state schools in the municipality of Piratini/RS. In other words, our audience includes 42 (forty-two) students from this school, who come from the countryside or the outskirts of the city - as long as they have some identification and/or closeness to the rural theme - and 24 (twenty-four) teachers from the most diverse areas of the municipal and state education system, as well as 1 (one) agricultural technician from the municipal garden.

1.4.2 Methodological contributions

In view of this study's intention to identify teaching strategies that provide students with an effective way of learning about, in this specific case, the biological role and reproduction of oligochaetes, we adopted a methodological itinerary that will be discussed in depth below. It should be noted that this itinerary, to a large extent, aims to highlight, simultaneously, "the theory of the approach (the method), the instruments for operationalizing knowledge (the techniques) and the creativity of the researcher (his experience, his personal capacity and his strong sensitivity)" (MINAYO, 1994).

With regard to the essentiality of method and techniques for the construction of scientific knowledge, Minayo (1994) warns that "as we are not geniuses, we need parameters" to be more precise. However, in order for this precision to be expressive, it is essential to combine method and techniques with "our stamp", our mark, registered through "our experience, intuition, capacity for communication and inquiry in any research work" (MINAYO, 1994).

In this respect, we used a qualitative and quantitative methodology which, although we are aware of the limitations imposed on the use of both, seem to us to be more in line with the research objectives. According to Prodonav (2013), the main differences between the two methodologies come mainly from the objects analyzed. While in the qualitative approach "the natural environment is a direct source for collecting data, interpreting phenomena and attributing meanings", the quantitative approach "requires the use of statistical resources and techniques, seeking to translate the knowledge generated by the researcher into numbers".

That said, we believe that our immersion in the field will be better presented in its stages of execution, which took place between October 20 and December 3, 2015. To this end, we will mention and comment on each of them below.

After contacting the management team at the Alaôr Tarouco Municipal Agricultural School, as well as receiving permission for the research from this team, teachers and parents, we scheduled our activities. We sent invitations to take part in our activities to students in the 6th to 9th grades of primary and agricultural education, to the school's teachers and others from the municipal and state education system in the municipality of Piratini/RS, to the agricultural technician from the municipal garden, and to the team responsible for filming the documentary - which we'll discuss below - in order to introduce the subject and the aim of the research and mediate our first analysis of the research.

Our research was therefore planned and carried out according to the following roadmap:

a) After the theoretical lessons on the biological role and reproduction of oligochaetes[11] , we administered semi-structured questionnaires (APPENDIX 01), i.e. with open and closed questions, to students enrolled in primary and agricultural education (grades 6-9). The purpose of the questionnaires was to ascertain their preconceptions about topics such as vermicomposting, the biological role and reproduction of oligochaetes, as well as to diagnose their interest in practical activities (Fig. 03A);

b) When we went to the school to administer the questionnaires to the students, we took the opportunity to schedule a next visit, during which another questionnaire, also semi-structured, was administered to the teachers (APPENDIX 02) at the school and at other schools in the municipal and state education networks in the municipality - for the latter, we went to their workplaces to administer the questionnaire. The teachers were given questionnaires with six multiple-choice questions and one essay, referring to the teaching practice used to teach the content on the biological role and reproduction of oligochaetes, as well as questions about the didactics applied, as well as their opinions on the practical activities;

c) Afterwards, we held a training session (Fig. 03B) to prepare and proceed

11 It is important to emphasize that the idea of investigating teaching from the perspective of theory linked to practice was facilitated by the working logic of the Alaôr Tarouco Municipal Agricultural Primary School. In other words, as this school is also involved in agricultural education and has a municipal vegetable garden within its premises, our research proposal was given expression when the teachers were willing to give theoretical lessons on the biological role and reproduction of oligochaetes - a situation that occurs on a daily basis in their classes - and, after which, with the help of the garden's agricultural technician, they collectively built a worm bin to practice vermicomposting.

with the construction of a worm farm on the premises of the Alaôr Tarouco Municipal Agricultural and Primary School, with the aim of combining the theory learned in the classroom with practice, in this case on vermicomposting, together with the students and teachers who answered the questionnaires, and with the help of the agricultural technician from the Municipal Garden (Fig. 03C);

Figure 03 - A) Detail of the student filling in the semi-structured questionnaire; B) Moment of training with the team in charge of the research, students, teachers and agricultural technician; and, C) Practicing the construction of vermicomposting worms.

Source: author's archives (2015).

Regarding the practice of building the worm bin for the vermicomposting process, it's worth noting that its aim was to explain to the students both the biological role - through the transformation of organic waste, in this case fruit from the school lunch (guava, orange and papaya), into humus, used as organic fertilizer, including for growing food - and the reproduction of oligoquetes. For this reason, our proposal focused more on proposing alternatives for integrating theory into practice than on using a tool to improve soil quality and, therefore, food production .[12]

The vermicomposting wormery was built using 12 (twelve) circular plastic containers without necks - pet bottles for mineral water - with perforated bottoms and a capacity of 5 (five) liters. Non-woven fabric (TNT) protection was placed on the top of the containers, attached to the edges with elastic to prevent the earthworms from escaping, as well as allowing for good aeration.

After preparing the plastic containers, 2,300 kg of dairy cattle manure was weighed and added. A portion of 10% was replaced by fruit waste from school lunch leftovers, depending on the treatment. There were a total of four treatments, with three replications each, using dairy manure and whole ripe fruit: guava, orange or papaya. The ripe fruit was used in its entirety, with peel and seeds, and cut into pieces to make it easier to distribute in the manure. The treatments consisted of:

- Tl= control (manure only);

- T2= manure + mulch;
- T3=esterco+goiaba;
- T4=esterk+orange.

Twenty (20) adult clitellate oligarchs of the *Eisenia foetida* species were placed in each bottle, along with the leiteiro cattle manure and fruit waste, according to the treatments above. A layer of rice straw was used to cover the bottles.

Throughout the experimental period of approximately 45 days, the containers were kept in a room/environment devoid of light and placed on a bench. During this period, as we will explain below, the students carried out evaluations in order to observe and identify, for example, possible earthworm escapes, a rise or fall in the humidity of the vermicompost system, and a smell characteristic of the evolution of the preparation. At the end of the 45 days, the number of

12 For more information on building a worm farm and vermicomposting, see, among others, materials from Embrapa (2016).

young oligocha, clitellate/adult oligocha and cocoons found in the system, i.e. in each of the treatments (T) of the practical activity, was analyzed.

d) It is important to emphasize that the practice of building the vermicomposting worm bin had the uninterrupted participation of a sound and video team, which filmed every moment of student and teacher interaction in order to produce a documentary at the end of the process ;[13]

e) Following the practice, we returned to the school to assess, together with the students, the relevance of building the vermicomposting worm bin. At this meeting, they prepared class reports in which they reported on the step-by-step practice and its relationship to the theory they had learned in class about the biological role and reproduction of oligoquetes.

As we mentioned above, although the practice of building the vermicomposting worm bin was carried out, it wasn't finished in itself. In fact, after its completion, students[14] and teachers divided into four groups, which they named: GROUP 1) The creepy crawlies, GROUP 2) Growing in Science, GROUP 3) Studying Oligoquetes and GROUP 4) Organic Compounds.

The students and teachers were divided into different groups based on the need to continue the practical activity. Knowing that 45 days after the practical activity of building the worm bin, the humus[15] would be ready, it was decided that, during this period, each of the groups would be responsible for a particular treatment (vermicompost from guava, orange, papaya and neutral waste[16]) and would analyze it on a weekly basis, passing on the information obtained to the other groups, analyzing the humidity, the galleries and the turning over of the soil.

13 At the end of the documentary, there are some testimonies from students and teachers about the practice.
14 The students were divided into 2 groups of 11 students and 2 groups of 10 students.
15 It should be noted that the hùmus produced in this didactic strategy was applied in the school's nursery and vegetable garden.
16 The neutral test refers to the vermicomposting process without the presence of fruit waste, in which case it was prepared using only dairy manure waste.

All the results obtained in the research were later analyzed and grouped into analytical categories in order to facilitate their appreciation for the discussion of the results. As a way of facilitating our research proposal and presenting the information gathered in the field with our empirical universe, the following image (Fig. 04) was drawn up together with the students and teachers in the school's computer lab, with the help of a graphic designer.

Figure 04 - Results analysis diagram.

Source: prepared by the author (2016).

The image above shows that our field of research is, to a certain extent, the space of the Alaôr Tarouco Municipal Agricultural Primary School. In this space, we understand that, because the teaching and learning proposal integrates primary and agricultural education, there is a significant presence of students, teachers and staff during working hours. In view of this, we need to consider that these people eat in the school canteen, consuming both natural and industrialized products.

We have already discussed the fact that a lot of the organic waste generated, to a large extent, is disposed of inappropriately in the environment, contaminating water sources and encouraging the development of a multitude of diseases. In order to minimize the harmful effects of the improper disposal of such waste, the school in question opted to find alternatives to disposing of it and accepted the invitation to take part in our project.

This led to the suggestion of adding fruit waste from school meals to the didactic practices of building a vermicomposting worm bin and then obtaining humus, taught by the teachers and attended by the students. Thus, based on the theoretical discussions in class about the improper disposal of domestic organic waste and the harmful effects it has on the environment, as well as the search for alternatives to minimize these effects, as is the case with the biological role and reproduction of oligochaetes (earthworms), is that students and teachers, mediated by professionals from the Municipal Garden, extend their discussions to practice, to know-how which, in this case, led to the construction of a vermicomposting worm bin. Remembering that the humus from this process has returned to the school environment, through its application in flowerbeds, in order to qualify food production in this space.

1.5 Work organization

After this brief introduction, in which we contextualize the themes that guide the research, such as identifying the problem, defining the objectives and hypotheses and constructing our methodological contributions, it is imperative that we continue with the clarifications relating to this dissertation. To this end, in Chapter 2 we will look at the theoretical references that have guided us. In Chapter 3, we will first outline the results of our research which, at the end, will be highlighted in the construction of our final considerations and proposals for future action, outlined in Chapter 4. Finally, we will show some of the evidence of our immersion and fieldwork, such as the interview script.

CHAPTER 2 SCIENCE TEACHING AND THE STUDY OF OLIGOCHAETES

"It is unlikely that any animal has played such an important role in the history of our planet as these little creatures (DARWIN, Charles)".

2.1 Initial considerations

This chapter brings out the main theoretical references that underpin the research. It's not just a question of commenting on the speeches and views of certain authors, but of revealing and discussing what is essential for understanding the work. While at first (Chapter 1 - Introduction) we set out to contextualize the dissertation, in this one we will look at the theoretical framework that underpinned our research choices.

To this end, the themes relating to science teaching will be discussed first. The idea of this section is to demonstrate how teaching, through practical activities, strengthens and corroborates the information learned by students during their theoretical training.

After a general overview of the subject of education, the processes of training and teaching for learning about oligoquetes will be discussed. The purpose is to demonstrate how fruitful the interaction between students and teachers can be, especially when the situations learned in the classroom are visualized in the subjects' everyday lives, i.e. through the practical demonstration of the transformation of organic household waste into organic fertilizer, humus - vermicomposting.

2.2 Understanding Teaching

First of all, it's worth pointing out that the word "teaching" has, like other words, "marks of its history, of its origin" (BARBOSA-LIMA, CASTRO and ARAÙJO, 2006). From this perspective, and following the thinking of authors such as Barbosa-Lima, Castro and Araujo (2006), we are led to understand "teaching" as the "art of transmitting knowledge". In order to understand the logic of the authors' reasoning, it is valid to bring into the body of the text some of their considerations about this word, namely teaching. According to them, when quoting Cambi (1999), the word teaching means

> It appears in the 11th century in French (*enseigner*), and in the 13th century in Portuguese; from the Low Latin *insignare*, an alteration of *insignire*, which means to indicate, to designate, and which can be understood as "to mark with a sign". Note that to mark can convey an idea of imposition, but on the other hand, it can indicate the birth of the student's character, since in Greek, sign is *Karakter*. The modern meaning of teaching (transmitting knowledge to a student) appears in the 17th century and, in fact, reflects the emergence of the modern school and the organization of the school system (CAMBI, 1999, *apud* BARBOSA-LIMA, CASTRO and ARAÙJO, 2006 - *authors' highlights*).

In view of the above, it is understood that the construction of knowledge, since ancient times, has been rooted, to a large extent, in processes that focus on the transmission of information from those who have it, to those who, at first, don't have it - as we will discuss below. However, as the term "education" has evolved over time in different societies, as well as the very re-signification of "teaching", significant changes have taken place in learning systems. As Barbosa-Lima, Castro and Araujo (2006) explain, education,

> in Portuguese and Castilian, is recorded in the 17th century. It appeared in French in the 14th century (*éduquer*), but was rarely used before the 18th century, and was poorly received until the 19th century. As recently as 1900, the *Dictionnaire Général* described it as a popular word. From the Latin *educare*, which is a derivative of *educere* and contains the idea of leading. From the same root come: produce, seduce, deduce, induce, translate, and others. The word educate represents a *praxis* in which the purpose and objectives of the pedagogical process are emphatically focused (*emphasis added*).

From this excerpt, we can see that the term "education" actually goes beyond the simple transfer of knowledge inherent in teaching. However, as Paulo Freire warns us in his book entitled "Pedagogy of Autonomy", and also thinking about the possible transformations of teaching processes over time, we can understand that this, teaching, is not limited to its original etymology - "the art of transmitting knowledge" (BARBOSA-LIMA, CASTRO and ARAÙJO, 2006) - but goes beyond it, creating "possibilities for its production or construction" (FREIRE, 1996).

It should also be noted that the noun "teaching" appears, in the words of Barbosa-Lima, Castro and Araujo (2006), "mainly in syntagms such as, for example, Physics Teaching", or Science Teaching. This is largely the case in this dissertation.

Notwithstanding the considerations made about teaching, in order to better understand its discussions, it is important to clarify and understand its origins, going through the different historical moments until we reach the current period, which is the proposal outlined below. Afterwards, we will bring "Science Teaching" into the scope of the study, from the point of view of the relevant processes.

2.2.1 A brief history of the teaching process

We can't deny that the attributes of teaching and education itself are directly related to the evolution of societies, and therefore of human thought. According to Cruz (2016), education is processed according to the specific socio-political factors of each era, as well as, of course, the search for the man of knowledge. For this reason, the author emphasizes that it is up to men to deliberate changes, to choose which educational lines will be followed and pursued for effective development, or not.

For example, if we look at the aspects of education that originated in antiquity, we can easily see that, in Clemente's (2000) view, they were designed "to train the people [...] of those times". We refer here to antiquity because, as the author reminds us,

> [...] the transmission of knowledge began as soon as human beings stood up. What one generation learned, it passed on to the others. It was oral communication, it was observing what others were doing, and so the sciences and techniques grew (CLEMENTE, 2000).

At the time, there were no social classes, i.e. people had the same degree of importance within the tribes, differing only in their functions. Even so, the knowledge learned from observing life and nature was passed down from the oldest to the youngest. Everyone had the opportunity to learn the essential techniques for maintaining life, from fishing and farming to domestic chores. In a way, the children "imitated" their elders, learning the "know-how" of each "thing".

With the passage of time and, therefore, the evolution of societies, the processes related to education have taken on other colors. In fact, in Cruz's view (2016), these colors have an impact on religiosity, on the construction of a heroic being, capable of ruling, with different shades and possibilities of interpretation and action. Even so, in Egyptian societies - one of the first societies to unveil the "mysteries" of writing - education gained a certain expression, with this society creating spaces for instruction, especially political, military and religious.

The perspective of "liberating" education, to a large extent, comes from the pedagogy developed in Greek societies. Although this initially met military and political demands (Homeric period), it soon gained in breadth, expressing itself in more philosophical practices (Athenian period). With this, it is understood that this civilization not only enhanced teaching didactics by integrating man, body and spirit, but also by awakening philosophical thought (CRUZ, 2016).

Unlike the peoples of Greece, Roman society took little interest in the realm of human thought, in this case philosophy. But, on the other hand, it focused its teaching and education on practices and norms seen as legal - some of which, such as private law, are still used, to a certain extent, today[17] . The Roman teaching-learning process was largely based on the figure of the father, who was responsible for transmitting knowledge to his children and then handing them over to the state education system - it was the Roman state that "educated" the "good" citizens (CRUZ, 2016).

With the decline of Roman society, we became immersed in an educational format linked to the deliberations of the Church. Throughout the Middle Ages, teaching and learning processes aimed to mold subjects in accordance with

---

17 For more information, see JusBrasil (2016).

sacred teachings, to form Christian men. In Aranha's (1996) view, in this historical period of education, there was a reduction in Greco-Roman pedagogical principles and a gradual expansion of pedagogical actions linked to the figure and image of God. In the author's words, "there is, therefore, a model of man, an essence to be attained for the greater glory of God" (ARANHA, 1996).

At the end of the Middle Ages, with the fall of the "fiefdoms" and the reappearance of cities, this belief in man as "God" was transformed into a belief in man. In the words of Cruz (2016), "at the end of the Middle Ages, the theological values broke down and anthropocentric values emerged, highlighting man's more human values".

The following period, the Renaissance, is characterized by a growing cultural and artistic movement, which completely escapes the verbalist and rigid formats imposed by the Church on education in the medieval era. According to Cruz (2016),

> the renaissance censorship of medieval education was in relation to formal and bookish teachings, in other words, the child had no freedom to criticize, challenge or seek new forms of knowledge, becoming dependent on this education of reading, with no propensity to analyze studies.

Thus, the Renaissance conception envisioned the hierarchical organization of learning levels (training of children, young people and adults) through education processes. In addition, as Aranha (1996) points out, the aim was not only to transmit knowledge, but also to form people morally.

The author comments that, progressively to the Renaissance, the Enlightenment era was marked by a rethinking of pedagogical practices. If before the spaces destined for the training of subjects had a classist posture, in order to differentiate those who possessed knowledge from those who didn't, in this Enlightenment era, the formation of lay schools common to all subjects began.

From the 19th century onwards, the development of capitalism brought about significant changes in the educational perspective, both in terms of the format of education - universal, lay, free and compulsory schools - and in the pedagogical dynamics of teaching, in order to provide training that dialogued with the reality of the subjects - the social function of teaching (ARANHA, 1996). Based on these assumptions, Souza, Souza and Teixeira (2014) state that teaching procedures which until then had been based on memorization practices, where the student was seen as a passive and receptive subject in learning processes, began to emphasize what Cruz (2016) understands as the intellectual dimension of man - active subjects in the process of understanding knowledge.

In the midst of all this, the situation here in Brazil was not entirely different. In this respect, Cruz (2016) states that when colonization of Brazilian territory began in 1530, our education was mainly based on the catechization of the Indians and the instruction of the settlers who lived here. This model of education lasted for more than 200 (two hundred) years and was later replaced by the Enlightenment ideas of European education. As Cruz (2016) explains,

> Instead of the typical humanities course taught by the missionaries [...], the classes were changed to the Royal Class System. Under the command of the royal crown, Portuguese teachers organized teaching plans based on dispersed and fragmented studies. These studies were applied by teachers who were often unprepared in relation to the content and learning taught, far removed from their own educational context and completely unaware of the classes.

This scenario in Brazilian education was characterized by the fact that a significant portion of the population remained uninvolved in education - high illiteracy rates. And those who did seek training were sent abroad, outside the country. In the words of Menezes *et al.* (2002), "relegated to the background, the education of the people was left to the personal and political interests of the Sovereign in the exercise of his absolute power". Nor after Brazil's independence, in Cruz's (2016) view, was there little change in teaching and education here in the country.

It was only from the 1930s onwards that Brazilian education gained momentum and achieved some expression on the national scene. According to Cruz, it was from this period that education underwent significant reforms, such as the creation of the serial curriculum, compulsory attendance, educational cycles and the qualification to enter both. Since then, various changes have been proposed to the Brazilian education system, proposals which, to a large extent, have led to the current format of education in the country. However, as our proposed study focuses above all on science teaching, it will be dealt with in depth below.

2.2.2 Science teaching - some considerations

In a way, science teaching became an integral part of teaching and learning processes from the 19th century onwards. Until then, the education of subjects was centered on the formation of knowledge through subjects more closely linked to philosophy, classical languages and mathematics. However, despite its brevity, science teaching was the target of different lines of thought regarding its appropriation. In the view of Chassot and Oliveira (1998),

> There were those who advocated science that would help solve practical everyday problems. Others focused on academic science, defending the idea that science teaching would help recruit future scientists. The second view ended up prevailing and although this original tension is still reflected in current science teaching, it has remained quite formal, still based on teaching definitions, deductions, equations and experiments whose results are known in advance.

From this point of view, the science area consisted mainly of the subjects of physics and chemistry. The subject of biology was only incorporated into science teaching later, given its uncertainties and high complexity.

With technological developments in the sciences, especially after the Cold War (1960s in the United States), a significant amount of resources was directed towards "thinking" about science teaching in "primary" and "secondary" schools, which are now elementary and secondary schools. The idea behind this theoretical and practical grounding of the sciences in formal education was based on the possibility of training "young scientists" for the future. Furthermore, in view of the neutrality of the sciences, which exempted its researchers from any judgments, its primary proposal was to

> develop rationality, the ability to make controlled observations, prepare and analyze statistics, and respect the requirement for replicability of experiments. In the 1950-70 period, the idea prevailed that there was a fixed and basic sequence of behaviors, which would characterize the "scientific method" in the identification of problems, the elaboration of hypotheses and the experimental verification of these hypotheses, which would make it possible to reach a conclusion and raise new questions (CHASSOT and OLIVEIRA, 1998).

This remarkable period in science teaching had a significant influence on different curricular trends. However, while before this teaching was strictly centered on the sciences, based on scientific neutrality, over the last few decades it has taken on board the changes that have taken place in the political, economic and social landscape. In other words,

> Once the idea of producing teaching projects had been overcome, in the 1970s we began to see the huge gaps in scientific training and in the general education of new generations in the face of the ever-increasing and rapidly changing need for knowledge. Criticism of the concept of science as neutral led to a new philosophy and sociology that began to recognize the limitations, responsibilities and complicities of scientists, focusing on science and technology (S&T) as social processes (CHASSOT and OLIVEIRA, 1998).

As a result, science teaching also began to take into account aspects of society, through its political, economic and cultural conceptions. To this end, the subjects who sought information and knowledge, namely the students, finally appropriated scientific content that was actually relevant to their daily lives. This enabled them not only to identify the problems present in their surroundings, but also to seek the best solutions.

New demands for information are emerging from old problems such as increasing pollution, waste, energy sources, natural resources, population growth, among others. Nowadays, it is not enough to teach isolated disciplines, but rather interconnected ones, based on interdisciplinary proposals.

In view of the above, there is no longer any way of perpetuating familiar teaching tools. In fact, we need to build new proposals for approaching the different themes, rediscussing the organization, including school organization

- proposals which are discussed below.

2.3 The Student at the Center of the Science Teaching-Learning Process

Humanity is currently going through a period marked by major transformations, mainly due to technological advances - which has a direct impact on Brazilian classrooms. In this respect, Antunes (2013) asserts that it is increasingly necessary to reflect on school attitudes in order to find solutions that improve the teaching-learning process, making it more attractive and meaningful. In the meantime, the student becomes a central figure in the teaching process, actively participating and using all the artifacts that surround them, as well as understanding theory and practice.

Notwithstanding this information, Fourez (2003) states that, in a significant proportion of classrooms, it is very difficult for students to learn certain subjects. There are many reasons for this, but at the end of the journey we will certainly see that it is largely due to problems in the application of teaching and learning processes. In order to improve these processes, the author emphasizes that

> The need to make school science content meaningful, as well as to discuss the role of science and technology in contemporary society, has become one of the most important issues on the educational scene in the last two decades (FOUREZ, 2003).

As Moran (2000) warns us, the problems in the teaching-learning process do not always come from the students, since teachers are an important part of it. There are cases in which the lessons given, because they are labeled as insignificant and unattractive, hardly get the students' attention.

There is no doubt that educational practices require renewal, both on the part of the student and the educator. To this end, it is imperative to review the tools used, especially by teachers, in the teaching-learning process.

The teacher is the one who has to direct the content in a critical and reflective way, using interesting approaches that are geared to the students' reality. It is therefore necessary to take into account the students' previous knowledge and experiences in order to find teaching strategies that effectively arouse their interest, as we will discuss below.

2.3.1 Reversing roles, changing importance

As we mentioned earlier, teaching and learning processes were often based on the practice of teachers transferring information to students who memorized it. However, between the nineteenth and twentieth centuries, what we know worldwide as the "New School" emerged, a transformative movement from a time of renewal. The New School emerged and became popular through new thinking, questioning conventional schooling, which was shown in the works of John Dewey (1859-1952), Jean Piaget (1896-1980), Paulo Freire (1921-1997) and many others. In the view of Antunes (2013),

> The New School changed concepts and criticized some of the pedagogical practices used by conventional schools, while remaining certain that good teaching does not dispense with a solid body of knowledge, all of which is grouped into subjects that give rise to a curriculum and which, in order to be understood, require a curriculum organization that contains programs and evaluations.

In this regard, Antunes (2013) asserts that there are nine (09) fundamentals of the "New School" movement which, in a way, have been promulgated and solidified over the years in order to differentiate teachers from each other. With regard to these fundamentals, the author points out that:

1) the autonomy of the student and the encouragement of personal initiative are the central pillars of quality education;

2) students are different and therefore learn autonomously and differently, according to styles that are never generalized;

3) the student's natural curiosity is the focus of their interest, around which learning situations are defined;

4) the student is the protagonist, proposing their active participation in the management of the learning and cooperation process;

5) the passivity of the student is opposed to their learning and, therefore, the valorization of methods centered on the interests and physical, cognitive, emotional and social needs of the students;

6) valuing outdoor activities and understanding that classroom learning is associated with experiments carried out in workshops and laboratories, practical and field lessons and many other everyday activities in the environment that surrounds them;

7) the articulation between intellectual life and the development of skills that do not dispense with material and social activities, valuing the dignity of all forms of work;

8) education is the essential means of enabling children to develop their personal skills and competencies to the full;

9) Punishment and any act aimed at curtailing the student's spontaneity must be banned.

For Antunes (2013), these foundations inspire subjects not only to reflect on teaching-learning practices, but also to direct their actions towards building more effective didactic strategies for classroom dynamics. In the author's words,

> The reflections that inspired the New School movement were based on the certainty that the teacher never teaches; in fact, they only contribute to the student's learning and that, in this way, a quality lesson should be based on the dynamic action of the student, led with subtlety by the teacher (ANTUNES, 2013).

In the view of Soussan (2003), the transformation of the old teaching logic - knowledge transfer - to more participatory formats is imminent, in other words, formats that encourage the confrontation of the information learned with the students' reality. From this perspective, Antunes (2013) asserts that, by enabling students to reflect on their personal experiences, with the consequent reconstruction of the signs and meanings of the realities that surround them, they will, at first, be empowered. Thus, the information worked on by the teacher will become knowledge for the student - the lesson taught by the teacher will aim to produce relevant learning.

However, in order for this to happen, it is imperative, according to Cordeiro

(2013), that teachers and students renegotiate the conditions of schooling, with meaning and importance, especially for the students' lives. Furthermore, this negotiation should generate significant changes in these students' conceptions and also in their sociability. In this way, the process of teaching will not, in our view, be an activity that is "done in a vacuum", strict to the content, because "teaching" implies adopting different procedures from those commonly used.

It's also worth pointing out that a real classroom isn't just about the space inside the classroom, but also about the physical environment (such as masonry buildings) and the spatial environment (which means life itself) it occupies - it's common sense that life, the world, is our best classroom, our best school. We teachers - and here I put myself in the context - are the "link" between some information and the students. We stimulate them to challenge themselves, to question themselves, us, others, etc., to expand their ideas. In short, we are the provocateurs who lead students to understand the knowledge, the information that is relevant to their reality.

We need to keep in mind that teaching is student-centered. They must be encouraged to ask and answer questions, contextualize and research, thus becoming the true protagonists of the teaching-learning process.

2.3.2 New Possibilities in Science Teaching

According to Carvalho (2000), science teaching in Brazilian schools is making increasing progress in training students to be questioners, researchers and aware of their relationship with the environment. However, it is still difficult for these same students to relate the theory proposed in class to the reality around them, often failing to distinguish the scientific knowledge present in everyday situations.

Faced with these issues, as teachers we have the great challenge of making science teaching even more interactive, enjoyable and exploratory. To this end, we should make use of some of the concepts of the "New School". As Tavares (2004) warns us,

Learning science is a challenge and should allow us to explore the environment in which we live. Knowledge will not be acquired just by reading or listening to the teacher, but by acting on the environment based on the students' curiosity and interest: observing, describing, investigating. Learning should be a stimulating adventure that leads to understanding that scientific research is not just a collection of disconnected facts, but everything that is around us, everything is life, everything is science. You just have to start looking with a scientific eye.

Teaching science, to a large extent, means innovating, transforming the focus of class dynamics. Students build their own knowledge, with the teacher mediating the process. In this logic, the teacher must work in a diversified, stimulating way and with current and relevant issues for social well-being, always aiming to commit the topics discussed to the quality of life of all subjects. According to Carvalho (2004),

This teaching proposal should be such that it leads students to construct their conceptual content by participating in the construction process and giving them the opportunity to learn to argue and exercise reason, rather than providing them with definitive answers or imposing their own points of view by transmitting a closed view of the sciences.

The excerpt above largely corroborates the themes discussed by Astolfi and Develay (1990). According to the authors, the different teaching strategies are expressively addressed in the growing research in Science Teaching.

Thus, it is understood that there are proposals on the horizon in Science Teaching that aim to replace the traditional model of lectures, modeled on the use of textbooks, with activities such as interviews, projects, practical activities, pre-tests, surveys, case studies, symposiums, experiments, problem situations, among others. As Astolfi and Develay (1990) make clear, these activities are viable for students' broad learning, because they use practical activities to create didactic strategies for classes based not only on theory. According to Luneta (1991),

Practical lessons can help the process of interaction and the development of scientific concepts, as well as allowing students to learn how to objectively approach their world and how to develop solutions to complex problems.

Nóvoa (1991) argues that the teaching format conceived here as alternative and dynamic makes the teacher available not as a mere teacher of content, but as a mediator of learning, a facilitator of the acquisition of certain concepts by the students. To this end, the teacher uses a variety of didactic strategies in order to make the student an active spectator, with high levels of motivation and learning.

In this way, we believe that research and the construction of answers to at least common questions will become characteristics of the new generation of students. Furthermore, and in agreement with the assumptions of the New School, it seems to us that by incorporating didactic strategies that are participatory, interactive and close to reality, teachers will provide their students with opportunities to become responsible social beings, autonomous and co-owners of their culture and knowledge.

But how to proceed? How can we change educational cultures perceived here as "hard" and internalized by societies? In order to demonstrate the feasibility of these possibilities, we will continue with the teaching strategies used for the theme of oligochaetes (earthworms).

2.4 Science Teaching and the Study of Oligochaetes

In this section, we propose to discuss the theoretical issues that relate to our research proposal, which focuses on the Science Teaching developed by the teachers at the Alaôr Tarouco Municipal Agricultural Primary School who, to a large extent, spend time in their classes teaching content related to the study of oligochaetes. To this end, it is imperative that we emphasize the essentiality of this study in Science Teaching, so that we can talk about the specificities of this knowledge for the student community.

2.4.1 Why study Oligoquetas?

A certain amount of disbelief and even ignorance about earthworms is common in today's societies. Many consider them to be nothing more than worms, while others think of them as fishing bait. However, those who experience the assumptions of food production, who see increases in ecological issues, those who are closer to the area of health, pharmacology, for example, would see them from other perspectives.

As Marques (2009) explains, earthworms have been widely used over the years to transform organic waste, such as household garbage and human waste, into odorless fertilizers used in food production. What would otherwise be discarded in the environment becomes, through vermicomposting processes - which we'll explain below - indispensable for maintaining soil quality and the growth and development of plant species. Furthermore, due to the wide variety of species found among earthworms (great biological diversity), Oppermann (2004, *apud* Marques, 2009) points out that

> Japanese scientists have managed to isolate an enzyme called lumbrokinase from the earthworm Lumbricus *rubellus*, which is being tested in prostate cancer treatments. In Germany and Colombia, two other earthworm substances are being researched for the production of antibiotics. At the University of Colorado in the USA, the worms are being used as guinea pigs for research into ageing. Among the many uses for earthworms is their use in the preparation of medicines for asthma, bronchitis, impotence, skin diseases, rheumatism, etc. They contain substances such as tyrosine and lumbrofoebrine, both of which have medicinal effects, the latter being antihypertensive.

These organisms not only contribute to the formation of tunnels in the soil for the aeration of plant roots and the penetration of rainwater, but also turn waste into fertilizer and are used to make different medicines and even some cosmetics. They are also used in cooking, for example in China and Taiwan.

Given the merit of earthworms for the ecosystem and for societies, it is imperative that they are included in school curricula, enabling students not only to get to know them, but also, due to their ease of handling, to visualize the theory in practice. In the words of Marques (2009),

> With regard to [...] composting, conditions could be created for the production of vegetables to be consumed in the school itself [...]. Over the course of the school year, conditions could be created for students to carry out this type of management on the school grounds. In this way, the school cafeteria could be involved in offering leftover vegetables, which are thrown away every day, for vermicomposting. And the students in the classes involved in the project would be responsible for

monitoring and maintaining the garden's activities. Finally, our "earthworm friends" would contribute to the production of humus. This would be a way of expanding horticulture in schools, as well as in science and biology classes [...].

The excerpt above shows, in a way, what we set out to investigate in this study. However, in order for our proposal to take shape in the discussion, especially of the results, it is necessary to inform the specificities that qualify earthworms as the "intestine of the earth" (Aristotle), and this is the outline of the discussion that follows.

2.4.2 Getting to know Oligoquetes

Oligochaetes are important soil organisms that belong to the Eukarya Domain, Animalia Kingdom, Annelida Phylum, Clitellata Class, Oligochaeta Subclass (RUPPERT *et al.*, 2005; MADIGAN *et al.*, 2010). It is believed that earthworms appeared on Earth approximately 570 million years ago, in the Edicarian period (Paleozoic Era), being one of the first organisms to appear on the planet (BOUCHÉ, 1983). After a long period of adaptation and diversification, Righi (1999) emphasizes that, from the ocean, earthworms migrated to the most diverse regions and are currently found in practically all terrestrial habitats, with the exception of glaciers, deserts, very acidic or salty soils.

According to Brown and James (2007), the vast majority of these essentially edaphic organisms live in the surface layers, between 30 and 50 cm of the soil profile, where they perform important ecological and environmental functions. Brown and Dominguez (2010) point out that, due to the intense movement of soil that they promote in the edaphic system, earthworms, along with termites, beetles and ants, are known as the "engineers of the ecosystem" because, as Davidson and Grieve (2006) explain, they transform and move the materials present in the organic and organo-mineral horizons of the soil, such as excrement from the meso- and macrofauna. Lee (1985) explains that,

In terms of nutritional category or feeding strategy, earthworms are divided into two large groups: detritivores and geophagous. Detritivores are those that feed close to the soil surface, ingesting mainly leaf litter, dead plant roots and other plant residues. Geophagous earthworms feed underground, ingesting organic remains with no recognizable cellular structure, dispersed among the mineral particles of the soil.

From this perspective, James (2000) states that the different ecological life patterns of the different types of earthworms present in the soil are reflected in the different functional roles and environmental effects of the species on the ecosystem. Therefore, knowledge of the diversity of earthworms in a given environment makes it possible to better understand the effects and changes that the activity of these organisms causes in the soil. Since our study focuses on species that transform organic waste into odorless fertilizers, we will emphasize the information on the *Eisenia foetida* species.

2.4.2.1 Eisenia foetida

Of European origin, the *Eisenia foetida* earthworm species is characterized by having yellowish or dark red rings along its body structure. Lourenço (2013) says that they "are wrongly known as the Califomian red worm" because, for the author, their real name is "organic waste worm" or "manure worm".

Because it feeds on semi-raw waste and is able to speed up the decomposition process, both by revolving the waste, favoring its aeration and homogenization, and by the chemical processing that takes place in its digestive tract, the earthworm species *Eisenia foetida* is one of the best known, acting above all in vermicomposting processes.

In this respect, Kiehl (1985) explains that the digestive tract of this earthworm promotes the assimilation of mineral compounds such as phosphorus, potassium and micronutrients. In its digestive tract, it transforms organic nitrogenous compounds into nitrates, as it has calcium glands that concentrate the calcium circulating in its organism. At the end of the digestive process, it produces a vermicompost with an average content of 13% organic matter. These droppings provide a very active organic fertilizer that promotes significant increases in the physical and chemical structure of the soil.

Given the biological importance of this species for different ecosystems, its reproductive role should also be mentioned. Knaper (1987), Hernândez (1991), Martinez (1995) and Ferruzzi (1996), among others, state that the *Eisenia foetida* earthworm species is the one with the best fecundity, longevity, body consistency, vividness of color, adaptation to captivity and low mobility - a fact that makes it suitable for vermicomposting.

According to Morselli *et al.* (1996), the reproductive capacity of this species is directly related to its high mating capacity throughout its existence. In this respect, Lourenço (2010) states that these organisms,

> Being incomplete hermaphrodites, it is necessary for two individuals to mate in order for the eggs of both to be fertilized, as there is no self-fertilization. Every seven or ten days, each worm produces a cocoon in the shape of a small grain from which its young later emerge. The incubation period can vary between ten and twenty days [...]. When they are first born, earthworms are white and turn color as their growth intensifies. What characterizes their puberty is the existence of the clitellum, a kind of protruding, light-coloured band around their body, located at 1/3 of their length (LOURENÇO, 2010).

We therefore agree with Morselli et al. (1996) that this species has been the most sought after and used in vermicomposting, not only because of its prolific qualities, but also because it adapts to the most different types of waste generated in spaces, including domestic waste. We will continue to discuss vermicomposting and its processes.

2.4.3 Vermicomposting

According to Compagnoni and Putzolu (1985), vermicomposting is a technique whereby different organic residues of animal or vegetable origin are subjected to fermentation processes associated with earthworms of the *Eisenia foetida* species for approximately 45 to 60 days. In the view of Bakker (1994), Tagliari (1997), Esteves (2000) and Nunes (2000),

> Vermicomposting as a process for obtaining vermicompost has emerged as a recommended practice for recycling organic waste. The vermicompost obtained can be used in the composition of substrates and as an alternative to minimize environmental impacts, which can be seen with the use of animal manure and food industry waste.

It is important to consider that the processes involved in vermicomposting differ significantly from conventional composting. According to Pereira Neto (1987), composting can be defined as a controlled aerobic process developed by a diverse population of microorganisms. According to the author, composting is carried out in two distinct phases, the first of which involves the most intense biochemical reactions, predominantly thermophilic, and the second, or maturation phase, involves the process of humification.

In composting, organic waste is transformed into vermicompost when it passes through the digestive tract of earthworms. In the words of Almeida (1991), Ferruzzi (1996), Bicca (1999), Hartenstein and Hartenstein (1981) *apud* Huber (2003), Albanel *et al.* (1988) *apud* Cabrera (2004),

> Vermicomposting differs from conventional composting, among other things, in that it forms humic substances more quickly by passing through the digestive tract of the earthworms used, favoring the appearance of native earthworms in the soil and improving its physical, chemical and biological properties.

Balota *et al.* (1998), *apud* Rodrigues *et al.* (2003), state that it is inside the digestive tract of earthworms that the material (organic waste) undergoes mutations and is subsequently transformed into organic matter to be made available to plants. According to Faganeli (2004, *apud* Marques, 2009), the contribution of earthworms and vermicomposting to the environment is

> The earthworm is a detritivore, which feeds on the detritus or organic remains of plants and animals. These, after being 'swallowed', are sent to the gizzard, where they are crushed and taken to the intestine for digestion. Eliminated through the anus, the worm's faeces are rich in food scraps that undergo the action of decomposer bacteria, thus fertilizing the soil. The humus (decomposing organic matter) is rich in nitrogen (N), phosphorus (P) and potassium (K), which are macro-nutrients needed by plants in general, among other components.

In this respect, Morselli (2005) states that the vermicomposting process promotes the formation of a material, vermicompost, which is richer in nutrients than that which comes from composting. Furthermore, as this material has a lower density and is in contact with earthworms, it reduces the amount of labor required in the process, as it does not require turning, excessive sieving or sieving to be used as soil fertilizer. Finally, vermicomposting allows the vermicomposting process to continue without the need to add more earthworms,

As well as vermicompost, the vermicomposting process offers alternative sources of income to the subjects, either by selling the humus, worms and/or feed, or by the possibility of obtaining a considerable quantity of humus and worms in small physical spaces. It also allows for proper recycling, by transforming "dirty" products (garbage) into "clean" products (odorless fertilizers) from waste that is usually treated inappropriately.

2.4.4 Organic Materials Used in Vermicomposting

According to Kiehl (1985), organic waste or decomposable material is a carbon-rich compound found in nature in the form of animal manure, food and vegetable waste (leaves, stalks, etc.), sewage sludge and others. According to the author, the fermentation process to which these residues are subjected causes the organic matter to decompose, resulting in a material called humified, stabilized or cured organic fertilizer.

From a biological point of view, manure is an excellent culture medium for microorganisms and also for the earthworms present in vermicomposting processes, increasing the amount of bacteria in the soil when used as a fertilizer. However, with regard to manure, it should be borne in mind that, according to Tibau (1985), it has the following compositions

> This varies depending on whether the animal species is ruminant or not, the age and physiological condition of the animal, the nature and quantity of the mass it receives as food, the type of handling and conservation of the manure and the composition of the bedding.

Still on this subject, Morselli (2005) explains that when manure comes from young animals, such as steers, the protein content of the material should be observed, as too much protein is harmful to earthworms. But in general, the author emphasizes that,

> In vermicomposting, any type of excrement can be used, but those from horses and sheep are more solid and much less watery than those from cattle and pigs, allowing for greater aeration, fermenting

more quickly, with greater heat development, unlike those from cattle and pigs, whose fermentation is slower.

To a large extent, Reinecke and Viljoen (1990) believe that cattle manure is an excellent source of food for *Eisenia foetida*. Because it is easy to find and easy to collect, cattle manure is considered, in the view of Ferruzzi (1996), to be a good substrate, an excellent food for earthworms, as well as for the production of humus. Despite the importance of cattle manure in vermicomposting processes, Morselli (2005) emphasizes that, apart from these,

> Virtually all other organic waste, whether vegetable or animal, can be used for vermicomposting, such as straw, sawdust, rice husks, fruit scraps, vegetables, food, leftovers from stables and animal bedding. In urban or industrial areas that generate organic waste, composting can be carried out using urban waste, sewage sludge, waste from slaughterhouses, canning factories, tanneries, waste from the manufacture of soluble coffee or coffee grounds, leftover vegetables and yerba mate.

As we can see, there are a multitude of materials that can be used for vermicomposting, for the production of vermicompost, or humus. Even so, instead of these materials being sent to vermicomposting, a significant portion of them are often disposed of inappropriately in the environment, degrading these spaces and promoting the emergence of countless infectious diseases.

In view of this, Rola and Silva (2014) consider that vermicomposting is one of the most viable alternatives for solving problems with organic and even industrial waste. Notwithstanding the viability of vermicomposting, it is important to pay attention to the fact that it should be given more space in the educational scene, in order to make people aware of the environmental obstacles caused by the improper disposal of waste and to point out successful alternatives that contribute to the life of the planet.

2.5 Final considerations

The above elaborations form our line of reasoning. In other words, they show the paths we have taken, the readings that have covered our theoretical framework. They start from a more general idea about teaching and education, going through the intricacies of Science Teaching to finally involve the study of oligochetes.

It wasn't easy to follow these paths, as each reading brought new information and new perspectives. Even so, we were able to dialog with different authors and build our theoretical framework, which, in our opinion, explains the next chapter, in which we will bring to the screen the main results of our immersion in the field.

CHAPTER 3 TEACHING SCIENCE: THEORY VS. PRACTICE

The purpose of this section is to present the main results of our research. Notwithstanding the information provided here on the results, it is worth emphasizing that they will be discussed and commented on in their own specific way. In this sense, and at first, the teachers who made up our research contingent will be portrayed. The idea of this subsection is to check the strategies they use to teach the content given in class. It is also a space dedicated to their reflections after experiencing the practical activity together with the students. Like the teachers, the students will also have a place in the discussions here, with their own section. In this section, they will present their pre- and post-conceptions about the content taught in class. In other words, they will talk about their knowledge of the theoretical lessons and, after taking part in the practical activity, they will talk again, demonstrating what they have learned. The purpose of these arguments is to express the relevance of practical activities for the subjects' training.

3.1 About the Teachers and their Teaching Practices

It is not new that teachers are seen by societies as responsible for the nature and quality of everyday educational action in classrooms and schools. In general, they are the ones who form and transform the mentalities of the subjects along more technical lines, in which the rules are derived from scientific knowledge, or autonomously, where the teacher makes decisions and creates in the midst of their own action (MENDES, 2011).

As Mendes (2011) points out, the concept of the teacher and the tools they use have changed over time, in order to keep up with the ideology of different theories. However, they have always favored learning, based on different approaches.

Although we recognize teachers' training processes, we know how difficult it is to take part in certain educational dynamics. We are often faced with complex realities, which lead us - and here I place myself as active in the teaching and learning process, as a teacher - to employ a set of strategies in order to deal with the content to be taught.

Our choices are based, above all, on the training we had as students, our everyday experiences and the wealth of information we collected individually along the way. For this reason, and in order to understand the tools used in the classroom by the teachers who make up our empirical universe, it is imperative to briefly survey their sphericities, such as where they teach, to whom, what their training is, among others, and this is the assertion that follows.

3.1.1 The Teacher Profile

As we mentioned earlier, the professional dynamics of teachers are linked to the social, political and economic circumstances of a given society. Furthermore, the passage of time influences the educational logic of these subjects, the teachers, because what was once widely used in classrooms may now be the target of revisions and even refutations.

For this reason, we must bear in mind that teaching is not rigid in itself. On the contrary, it has to be reflexive and malleable in order to adapt to different situations.

Even so, we still find many teachers who focus their work on certain territories, either because they prefer to work near their home, near their family, because of the ease of access to certain schools - they generally prioritize urban schools over rural ones - or because of the availability of time, among other reasons, such as the determination of the location of the vacancy, given by the public education institution (city hall or state). For example, in the scenario of the teachers who make up the empirical universe of this research, a significant proportion of them work in schools located in the urban area of the municipality of Piratini/RS (Fig. 05).

Figure 05 shows that of the 24 (twenty-four) teachers investigated in the study in question, which was carried out at the Alaôr Tarouco Municipal Agricultural

Primary School, located in a rural area, less than 17% of them work at the aforementioned school, i.e. 4 (four) teachers. The other 83%, or 20 teachers, teach in schools located in urban areas.

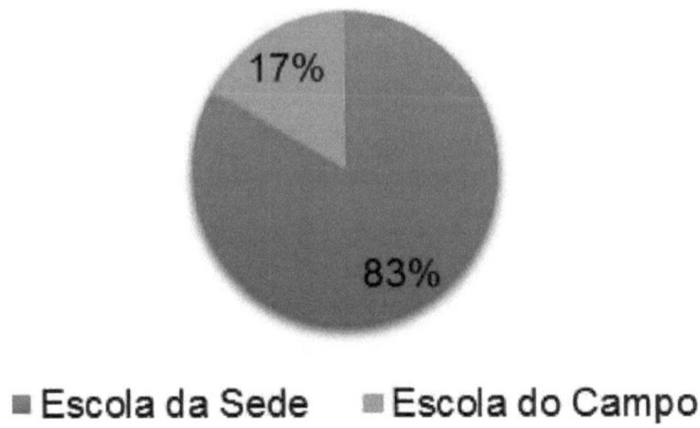

Figure 05 - Schools of work of the teachers who make up the empirical universe of the research.

At this point, it should be made clear that although the school has a teaching staff that meets its teaching needs, our research proposal goes beyond the confines of the school. For this reason, other teachers from the municipal and state schools were invited to take part in the research. This situation explains, to a large extent, the expressiveness of the teachers based in the local schools, that is, in urban schools, since there are approximately 49 (forty-nine) teachers in the Municipal Education Network alone in the areas of Languages, Mathematics, Science and Physical Education.

Another important piece of information refers to the type of teaching the teacher is involved in. In other words, apart from the subjects they teach, it is important to know which elementary school grades they work in. This results in the depth of the topics they cover for the different levels of students.

In our empirical universe, we can see that the teachers work significantly in the 6th (sixth) year of elementary school (Fig. 06). However, this does not mean that they do not work in other grades of primary school. According to Inep (2009), "it is very common to find teachers who teach two or more classes in the final years of elementary school". This situation is shown in figure 06.

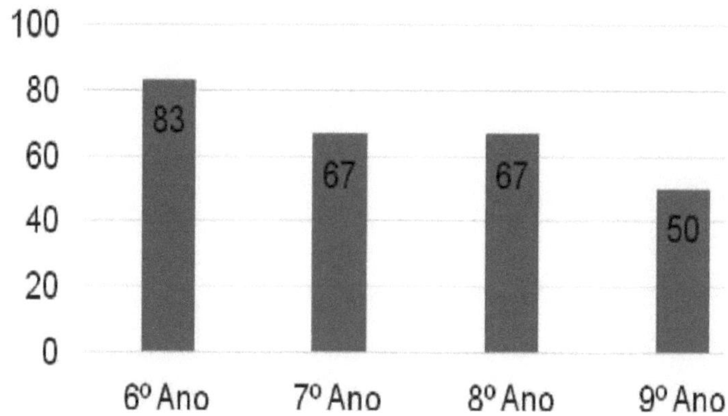

Figure 06 - Elementary school grades of the teachers who made up the empirical universe of the research.

It is interesting to note that half of these teachers have been working in the final grades of elementary school for between 4 and 10 years, and the other half have been working for more than 10 years in municipal and/or state schools. This difference is significant if we are to think that a more recent teaching career weighs heavily on this professional's accumulation of knowledge.

In other words, this teacher has had little experience of classroom theory in practical activities. This in no way disqualifies them as professionals. However, it does prove, for Paganini (2012), certain shortcomings in linking theoretical and practical knowledge, in formulating real experiences for students to appropriate this knowledge.

On the one hand, half of our sample is still in the initial stages of their teaching career, but on the other hand, the continuity of their training is appreciable. In other words, all of the 24 (twenty-four) teachers who took part in our research went on to complete their undergraduate degrees and postgraduate courses.

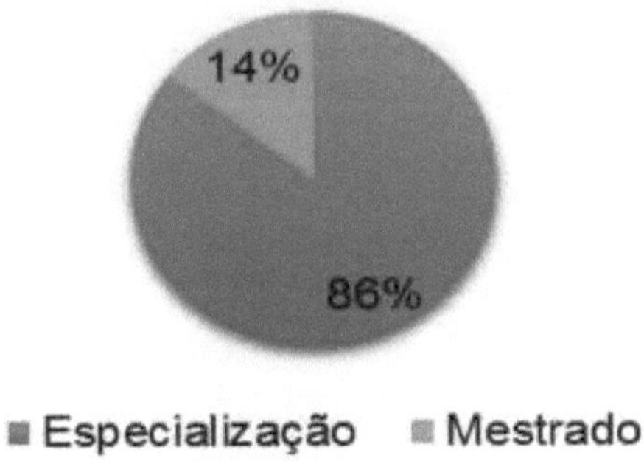

Figure 07 - Level of training of the teachers who make up the empirical universe of the research.

As can be seen in figure 07, all the teachers in the survey have post-graduate degrees, 86% of whom have a specialization degree and 14% a master's degree. For various reasons, such as lack of time, the need to be released from the teaching institution for professional training, among others, our interlocutors have not entered, or have not yet completed, doctoral courses.

As can be seen from the information presented above, the teaching scene in the research is significantly eclectic. If we also take into account the information provided in the documentary, we can see that the areas in which these teachers work are also different.

Teachers from the areas of science, Portuguese, mathematics, physical education and agricultural practices took part in our research, some with a specialization in Environmental Education and master's students in the area of Education and Technology, for example .[18]

Faced with such diversity, there is also a certain diversity in the teaching methodologies they use, which will be discussed in the following section.

3.1.2 Main Teaching Methodologies

Fourez (2003) and Antunes (2013) reminded us at the beginning of Chapter 2

18 It should be pointed out here that our invitation was extended to all the teachers in the municipal and state schools. However, only a few showed up on the scheduled date to carry out the activities involved in this dissertation.

of this dissertation of the need to make teaching and learning processes more attractive and meaningful. In the meantime, it is imperative that teachers reflect on their didactic dynamics, their teaching methodologies, in order to facilitate the appropriation of knowledge by students.

In this respect, Bergamo (2010) explains that teachers must build up the "investigative spirit" in students, which will awaken their curiosity about reality itself. But for this to happen naturally and efficiently, it is necessary to use methodologies that allow for a greater number of simulations in class. Also according to the author, teachers need to relate the content of textbooks to the present and the daily lives of students, so that the knowledge covered in class makes sense to the student.

In this respect, we asked ourselves: how will our interlocutors, the teachers, act in the classroom? What methodologies do they use?

The answers to these questions can be found in this subsection. However, before we continue with the discussion of the results, it is worth remembering that the information presented here summarizes the actions of 18 (eighteen) of the total 24 (twenty-four) interlocutors in specific classes dealing with the appearance and reproduction of oligoquetes, their biological role and vermicomposting .[19]

According to the information obtained in the questionnaires, we realized that all the teachers use texts from textbooks or scientific books to teach the content relating to the appearance and reproduction of oligochaetes (Fig. 08). However, this scenario does not corroborate the idea that they don't use other methodologies in their classroom dynamics.

---

19 In addition to this information, it should be noted that of the 24 (twenty-four) teachers who accepted our invitation and took part in the research activities, only 6 (six) were from areas other than science. In other words, they teach subjects in the areas of languages, mathematics and physical education. In this perspective, although they took part in the research, they did not answer the questions posed in the questionnaires aimed at collecting information about the methodologies used by the teachers in class on the emergence and reproduction of oligochaetes, their biological role and vermicomposting, since this content is not covered in their classes.

Figure 08 shows a certain parity between the use of practical activities and videos and/or documentaries as a complement to the use of textbooks. However, the use of other methodologies, such as group dynamics, seminars, etc., did not feature in our results.

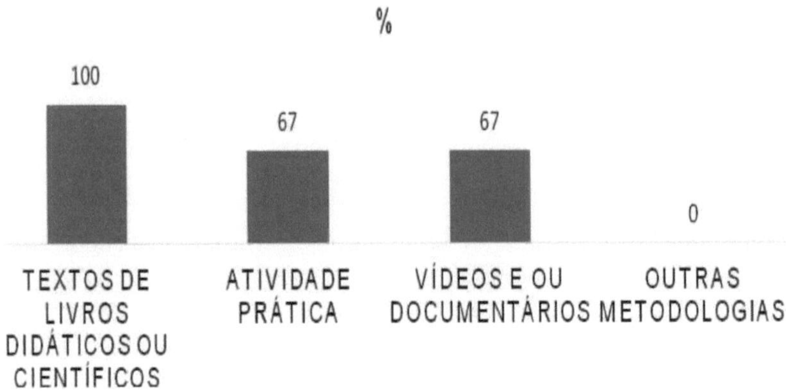

Figure 08 - Methodologies used by the teachers who make up the empirical universe of the research to explain the emergence and reproduction of oligochaetes.

A similar situation occurs with lessons on the biological role of oligochaetes. It is not uncommon for teachers to use texts from didactic or scientific books for teaching dynamics, leaving practical activities and the use of videos and/or documentaries as complementary mechanisms (Fig. 09).

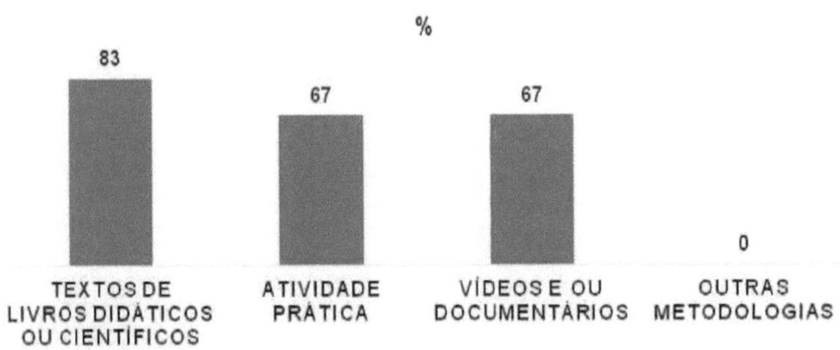

Figure 09 - Methodology used by the teachers who make up the empirical universe of the research to explain the biological role of oligochaetes.

Even so, it was possible to observe that teachers use more complementary strategies, represented here by hands-on activities and videos and/or documentaries, in classes where they teach content related to the biological role of oligochaetes, than in classes where they deal with their reproduction.

This may be due to the fact that the themes on the pages of textbooks and scientific books are associated with animal reproduction, including the reproduction of oligochaetes. The biological role of oligochaetes is often dealt with separately from the subject syllabus, a fact that encourages teachers to look for alternative didactic methodologies.

This assertion is backed up by asking about the tools used by teachers when the subject taught in class is vermicomposting (Fig. 10).

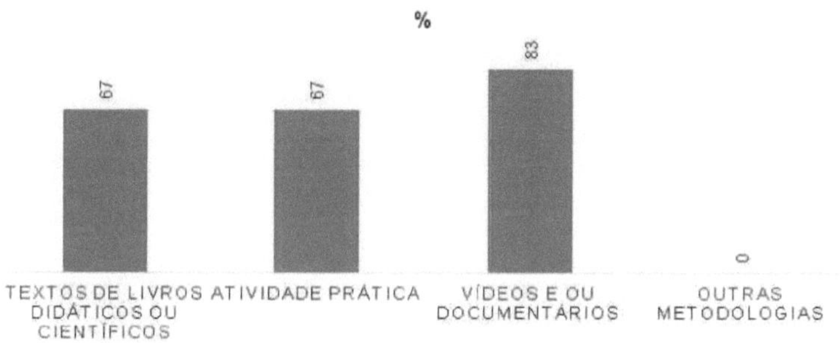

Figure 10 - Methodology used by the teachers who made up the empirical universe of the research to explain the content of vermicomposting.

According to figure 10, videos and/or documentaries are the mechanisms chosen by more than 80% of teachers to work on the subject of this type of lesson. It should be borne in mind in this context that many schools don't have the space to hold practical lessons on vermicomposting, leaving them to visualize the content through videos and/or documentaries.

The information detailed above shows that, although teachers are aware of the limitations of using theoretical didactic tools to help students grasp knowledge, they don't always "let go" of them when detailing information in class. However, even at a "slow pace", it is clear that they are moving towards dynamic, practical activities that involve the effective participation of the students.

3.1.3 From Theory to Practice - Some Considerations

Neves (2012) provides an interesting analysis in her work entitled "From theory to practice: a bridge to be built from initial training". In it, the author comments on the case of a teacher who, when faced with the reality of the classroom, came into conflict because he couldn't see any relationship "between the syllabus studied during his academic training and practice".

Taking this teacher's experience as a starting point, we oriented this sub-section towards uncovering some of the teachers' perceptions of the approximations and distances between the content taught in theoretical and practical moments. "Word clouds" were drawn up from the answers written on the questionnaires (and also from the students, as we'll see below), since they make the words repeated continuously in the speeches more prominent and colorful, giving them meaning and expression. This is how we assimilate the information obtained, identifying the most relevant expressions that synthesize the information.

For example, when we asked, "*Can we say that practical activities make it easier for students to grasp the proposed content?*", the teachers told us that yes, practical activities are very relevant in these circumstances (Fig. 11). As can be seen in figure 11, the words practice, student, theory and yes stand out from the others, demonstrating their importance in the answers given by our empirical universe.

Figure 11 - Word cloud[20] expressing teachers' views on the statement that practical activities make it easier for students to grasp the theoretical content they have seen in class.

It should be noted that these words, especially "practice", appear in the teachers' answers in connection with statements such as: "it's necessary" and "of course". In order to illustrate this point, some of the answers given in the questionnaires are included in the text:

Yes, because practice is necessary for the student to be able to understand the theory and thus build their knowledge and be able to explain it. / Absolutely, because once the student has the opportunity to put theoretical knowledge into practice, it becomes much easier to remember and associate theory and practice.

Figure 12 shows that *a* significant proportion of the teachers' answers to the question "*Do you believe that through practical activities it is possible to arouse the student's curiosity and, consequently, gain a greater understanding of theoretical knowledge? Why?"*, there is the word more, followed by practice, learning and student. This leads to the statement that the teachers who make up the empirical universe agree that practical activity enhances students' theoretical learning in class, since it motivates everyone to participate in the teaching and learning process.

20 The proposal to use the "word cloud" to analyze the descriptive content of the questionnaires applied to teachers and students comes from the idea that these "clouds give more prominence to the words that appear more often in the source text" (AFONSO *et al.*, 2011).

Figure 12 - Word cloud expressing the teachers' view of the statement that through practical activities it is possible to arouse the student's curiosity and consequently gain a greater understanding of theoretical knowledge.

In the words of the teachers to the question posed in the previous paragraph, there are the following answers that corroborate the information provided in the word cloud drawn in Image 12:

I think so, because through practical activities the students show more interest in the content, making it easier for them to learn. / Certainly, because everything that is experienced in practice becomes easier to assimilate and, in this way, learning becomes more exciting and enjoyable.

To a large extent, this information is in line with that obtained by Souza *et al.* (2014). The authors, based on studies with Science and Biology teachers from three public schools in the city of Altos, Piaui, point out that practical classes are perceived by teachers as the most appropriate strategies for student learning.

Despite the importance given to hands-on activities by the interlocutors, they were emphatic in stating that the content they teach in the classroom, especially about the reproduction of oligochaetes, is sufficient for students to understand (Fig. 13). Even so, they added that it was always necessary to look for new teaching strategies.

Figure 13 - Word cloud expressing the teachers' view that the methodology applied in their classroom to exemplify the content of the reproduction of oligochaetes (earthworms) is sufficient for student understanding.

The word cloud above is associated with the question: "*Do you believe that the methodology applied in your classroom to exemplify the content of reproduction of oligochaetes (earthworms) is sufficient for student understanding, or do you believe that you should renew it and look for new didactic strategies to apply? Which ones?*". The teachers answered this question:

The methodology presented was sufficient, but there is always a need to be open to new ideas. / I believe it is sufficient, but there is always a need to renew and look for new strategies to apply, because every year is different and there are always new students and new curiosities.

It is important to remember that, in the classes they teach on reproduction, the teaching strategies are mainly based on the use of texts from textbooks or scientific books, which give the class an eminently theoretical character. In these circumstances, it has to be considered that not all schools have the necessary tools for teachers to apply different methodologies in the classroom. This fact was constantly mentioned (orally) by them during the research, and was also mentioned by Souza *et al.* (2014). For the authors,

Even lectures, which are the most widely used, are often hampered in their efficiency by the lack of basic technological resources such as the data-show, which most of the time, due to the insufficient quantity [and when there is one in schools], there is a rotation as to how many times it can be used.

3.1.4 New Perspectives from Practice

After taking part in the training session we held, together with teachers from various schools and students from the Alaôr Tarouco Municipal Agricultural School, as well as the agricultural technician from the Municipal Garden, interesting things were said in the testimonies available in the documentary that bears the title of this dissertation. A significant proportion of the teachers who took part in the current investigation told us that they had encouraged their home schools not only to get students actively involved in practical lessons, but also to ask the school management for their own spaces to develop these activities. In the words of Professor Alexandra ,[21]

"I took part in Professor Mariana's training on vermicomposting, I was very interested, and I proposed that my class at the Repùblica Rio-Grandense school do it. We set up a compost heap [...] with the authorization of the school's management team. The students worked hard, the eighth grade class worked together, they were very dedicated to the process [...]. This work was rewarding because the training encouraged me to get my students to seek out and contribute to the environment. We used to have a lot of organic waste produced at the school that was thrown in the garbage can, but now it's not. Now this waste is kept in the compost garbage can, there's a vermicomposting process, which produces the compost and we put it in the vegetable garden. And the products produced in the school garden are consumed by the students themselves."

The excerpt expresses the importance of the practical activity carried out at the Alaôr Tarouco Municipal Agricultural Primary School, not only for the school's public, but also for others, such as teachers from the municipal and state education systems. According to the testimony above, the vermicomposting activity went beyond the reproductive logic and biological role of the oligoquetes, as it provided theoretical and practical elements for the development of projects in schools in the municipality of Piratini/RS.

In other words, the content covered in science teaching has led to essential

21 The names of the interlocutors appear in the dissertation and in the documentary because, in addition to their authorization for this measure, we believe in valuing their knowledge and their speeches for our work.

changes in the daily lives and experiences of students and teachers. Minhocarios were built with the intention of making it possible to develop practical lessons linked to the theoretical content seen in the classroom, and also to reuse school lunch waste to produce food for school lunches. For Professor Gabriel,

"When we see theory combined with practice, we realize how important it is for our students, for our community, that practical activities like this are carried out. What we've seen today is the realization of a worthwhile project and we can see that our students are able to take more meaningful learning home with them. I would also like to report on a proposal developed by the Deputado Adâo Preto State High School, where I work with the Integrated Seminar subject, and where we have developed an agro-ecological vegetable garden near the school. So, there we also have a compost bin, we also have a wormery that we managed to develop using the waste produced by the school, and we are already using the humus to produce vegetables in the school garden, which has been working very well. What we've learned here today will help us to optimize the work carried out at our school a little more, and to expand, in fact, this question of theory and practice, which is to enable our students to put everything they learn in theory, in the classroom, into practice and have a more significant result."

The training session, followed by the practical activity, was also very significant for the teachers at the Alaôr Tarouco Municipal Agricultural Primary School. In their view,

It's always important that, with the students, what they learn theoretically in the classroom, they try to show in practice. Together with my fellow teachers and the school's management and coordination, this is the pedagogical policy we adopt here. In the classroom, when we show the theory, we then go into practice, here in the municipal garden, which offers us the chance to bring the students into the possibilities, the conditions, to show them how what they've learned in the classroom is done. This movement between theory and practice is fundamental to education. This increasingly strengthens and improves the teaching-learning relationship (Professor Marcones).

I followed Professor Mariana's project, which was carried out weekly here at the school with the afternoon students, who are the ones who do the practical part. It was very useful for them. They showed a lot of interest, they took part every Thursday and, for them, it was something new, something different. They had the opportunity to improve their knowledge, right, by asking questions, clearing up doubts, and for us too, as teachers, it was something that contributed to our work, where we could associate theory with practice (Professor Marcelia).

In view of the above, there is no denying the viability of practical activities as a strategy for the appropriation of theoretical knowledge by students. Despite the difficulties, such as lack of space, equipment, availability of staff, among others, teachers who had previously dedicated themselves mainly to theoretical classes have sought to broaden their teaching methodologies, improving their knowledge and enabling students to take a different view of their education. As Professor Lúcia recalls, "it's important for us to put classroom theory into practice, and the students were very capable and happy to do the work".

3.2 Student Experiences and the Study of Oligoquetes

Recalling what Carvalho (2000) emphasizes about the questioning, participatory student - the centerpiece of the teaching and learning process - we will discuss the intricacies of the students at the Alaôr Tarouco Municipal Agricultural Elementary School.

As we explained in the methodological section of the research, the choice to focus on this school and its students is explained by its proposal to teach not only elementary school content, but also agricultural teaching, bringing scientific knowledge closer to the students' intrinsic reality. This situation was witnessed by us during the Vermicomposting practical activity.

In a way, vermicomposting meets current demands to reduce the improper disposal of organic household waste. What's more, by transforming this waste (from garbage to fertilizer), it enables significant improvements in the physical, chemical and biological quality of soils. With this in mind, the practice of vermicomposting is recurrent in science teaching within agricultural studies. Thus, we already knew that students attending primary and agricultural schools would have theoretical lessons on the reproduction and biological role of oligochaetes, with a subsequent practical activity on vermicomposting.

This led us to invite these students to take part in our empirical research. After accepting the invitation and taking part in a theory class on the subject, they answered a questionnaire. The answers they gave in the questionnaire formed what we call "pre-conceptions", since they represent the knowledge they acquired in the theory class.

After the theory lesson, the students did a practical activity[22] in order to improve their understanding of the reproduction and biological role of oligochaetes (earthworms) by building a vermicomposting bin. At the end of this practice, the students write a report on the lesson, specifying the operations carried out during the period. The reports they write form the basis of our research, as they conceive of theoretical knowledge enhanced by visualization and practical action.

In order to highlight the students' perspectives on knowledge and teaching methodologies in theoretical and practical classes, this subsection is divided into three (03) distinct moments. In the first, the students' preconceptions about the content of reproduction and the biological role of oligochaetes (earthworms) will be presented. In the second, their expectations regarding the practical activity of vermicomposting will be presented. And finally, in the third section, they will present the concepts they described in the report.

3.2.1 The know-how of students from the perspective of the classroom

The idea of entering the students' world is to understand their conceptions of the processes of teaching and learning through their worldviews (TEÓFILO and DIAS, 2009). Thus, when we asked our interlocutors, the students, about the theoretical content they had seen in class, they showed certain particularities that we will discuss in this material.

With regard to the information they learned after the lecture on the reproduction and biological role of oligochaetes, there was some knowledge, especially regarding the inclusion of earthworms in the Oligochaetes group[23] . As can be seen in figure 14, a significant number of students remembered the theoretical

---

22 The practical activity to which we are referring is in line with our research objective of producing a documentary. It was filmed during the dynamics of a practical lesson on vermicomposting processes, together with students and teachers from the Alaôr Tarouco Municipal Agricultural Primary School and other municipal and state schools in the municipality of Piratini/RS. This document will not be restricted to the research scope of this master's student, but will be distributed to schools in the region, as well as being made available on the public web, to facilitate its distribution and improve access.

23 For more information on this topic, see section 2.4.2 of this dissertation.

information provided in class by the teacher, that earthworms belonged to the *Oligochaeta* subclass.

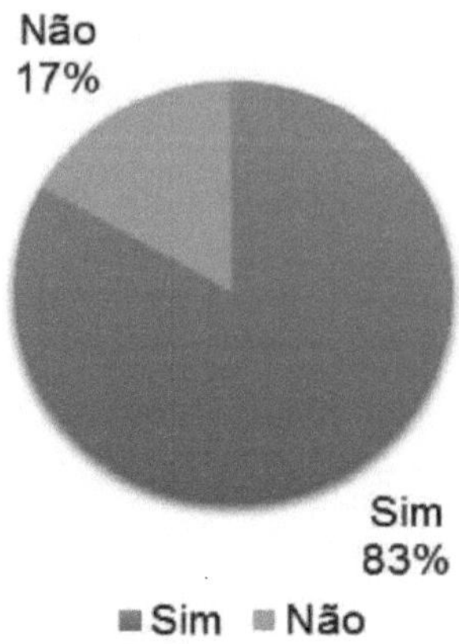

Figure 14 - Knowledge acquired by the students in theory class about earthworms being part of the Oligoquetas.

Furthermore, when asked about "*oligochaetes are incomplete hermaphroditic animals that require two individuals to mate in order for the eggs of both to be fertilized. Every 7 or 10 days, each worm produces a worm   in the shape of a small egg from which the young later emerge. What is the name of this little bump?*" The students answered the question correctly, i.e. they filled in the gap with the correct option (Fig. 15).

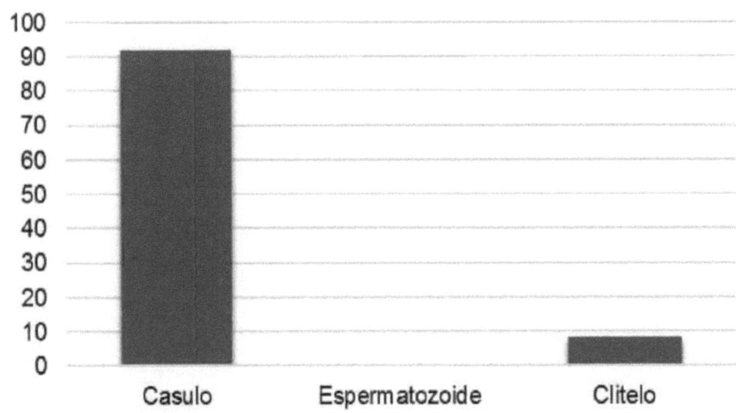

Figure 15 - Students' knowledge of oligochaete reproduction (where the young come from) in a lecture.

The appropriation of the knowledge seen in theory class by the students can also be seen from the information contained in figure 16. From their accounts of the biological role of earthworms in ecosystems, a word cloud has emerged, sparking words such as humus, fruit, manure, earth, waste, transform. This cluster of words draws our attention in order to reaffirm the students' understanding of the theoretical content.

Figure 16 - Word cloud expressing the knowledge acquired by students in lectures about oligochaetes being considered "ecosystem engineers" (perspectives on their biological role).

Among the answers given by the students to the question "*Oligochaetes are considered the engineers of ecosystems, do you know why? What is their biological role?*", from which the word cloud above (Fig 16) was drawn up, they gave us their answer:

Because they turn the earth into hùmus / Turn fruit and manure into hùmus / Turn waste into hùmus.

It is believed that the reaffirmation of the assimilation of theoretical knowledge by the students, as shown in figures 14, 15 and 16, is directly related to the period of their attention in the classroom. In other words, "in the search for new things that keep coming, attention constantly changes focus and is subject to exhaustion in fractions of a second" (DE-NARDIN and SORDI, 2008).

This fact gains meaning and expression when we turn our curiosity to the evaluation of the answers given by the students and analyzed in figure 18. In this figure, we can easily see how dispersed the students are in class.

Although figure 17 shows words like waste and humus, which are typical of vermicomposting processes, it doesn't answer the question asked in the questionnaire, which asked for the name of the process used to obtain humus. As can be seen in figure 17, the word composting comes up in the word cloud. However, as Pereira Neto (1987) points out, composting and vermicomposting are processes that differ significantly.

Figure 17 - Word cloud expressing the knowledge acquired by the students in theory class about the name and process of obtaining humus.

While composting transforms waste into organic compost through controlled aerobic fermentation with the help of micro-organisms. Vermicomposting uses the digestive tract of earthworms to transform organic waste into humus. It is true that both organic compost and humus are products that, when incorporated into the soil, increase its physical, chemical and biological quality. However, humus is only produced through vermicomposting processes.

Another question answered by the students that shows their disconnection from the theoretical content taught in class is shown in figure 18. Instead of referring to the "clitellum" to identify young earthworms, a significant proportion of the students attributed this identification to the spermatozoa.

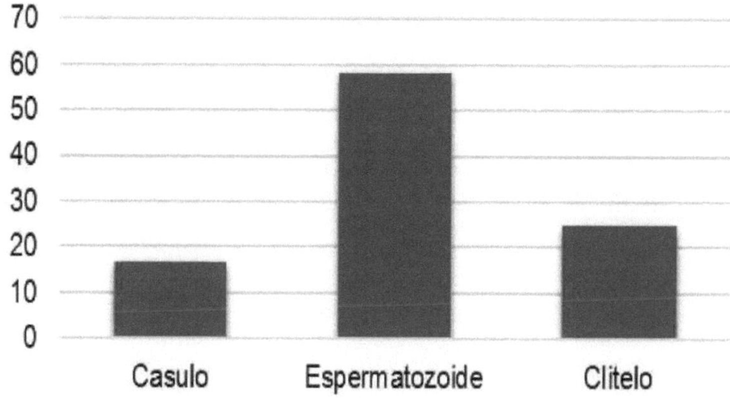

Figure 18 - Knowledge acquired by the students in theory class about the growth and development of oligochaetes (identification of adult earthworms).

In view of the information discussed above, it cannot be denied that the theoretical information given in the classroom is, in a way, far removed from the students' reality. For, as DeNardin and Sordi (2008) remind us, there comes a time during theory classes when students become dispersed and no longer pay attention to the content being taught. As Giroux (1986) said,

Theory and practice represent a certain alliance, not a unity in which one dissolves into the other [...] theory cannot be reduced to being perceived as sovereign over experience, capable of providing recipes for pedagogical practice.

3.2.2 Beyond the Classroom

If the content taught in theory classes is, according to the information discussed above, incompletely assimilated by the students, it is imperative that teachers look for alternatives, i.e. other methodological teaching strategies. But it is imperative to bear in mind that not only teachers think about the lessons they teach. Students also have their own expectations, which are presented below. As Bergamo (2010) points out,

"The teacher today is no longer the sole holder of knowledge, where only he speaks and the passive students listen or pretend to listen. The new role of the teacher in this current context is that of knowledge mediator. They need to create opportunities for their students to think for themselves, for ideas to be discussed, providing moments to review ideas, deconstruct hasty opinions by problematizing or proposing alternatives to overcome difficulties. In this process of intellectual autonomy, dialogue between teacher and student is very important."

In fact, for this author, it is necessary to use a framework of motivating methodologies that demonstrate humor, enthusiasm and practical application. Although for students these methodologies are largely divided into theoretical and practical, they don't require demonstrations of humor, enthusiasm and practical application.

The answers given by the students, the empirical universe of our investigation, to the question "*Do you believe that the contents explained by your teacher in science classes would be easier to understand if they were worked on through practical activities?*", allow us to understand that they do believe that it is easier to assimilate the theoretical contents with a subsequent practical activity (Fig. 19).

Figure 19 - Word cloud expressing the students' opinion as to whether or not they believe that the content explained by the teacher in science lessons would be easier to understand if it were taught through hands-on activities.

In figure 19, the question answered by the students comes from the question "*Do you believe that the contents explained by your teacher in science classes would be easier to understand if they were worked on through hands-on activities? Why?".* Among the answers they gave us were:

Yes, because it seems that we learn more in practice. / Yes, because then it would be easier, practice is cooler. / Yes, because we learn more by doing and looking.

In the above question, as shown in the word cloud (Fig. 19) and in the excerpt above, the words that stand out represent the students' worldviews on the subject of teaching, whether theoretical and/or practical, and are represented by expressions such as yes and more. In their view, the assimilation of content through theoretical methodologies linked to practice is more effective than using only theoretical methodologies.

The students' perspectives on taking part in practical activities, as a complement to the theoretical assimilation of the content taught in class, are expressed in their expectations. With a practical activity on the reproduction and biological role of oligochaetes (earthworms) on the horizon, based on vermicomposting processes, the students commented in the questionnaire on which organic waste would be preferred by the worms.

Among the possibilities mentioned in their answers (Fig. 20) were fruit waste such as guava, orange and papaya. It is worth noting that these fruits are widely consumed in school meals.

Figure 20 - Word cloud expressing the students' opinions about the practical activity involving the didactic unit Oligoquetes and organic waste, in which they emphasized the fruit that would be preferred by the worms.

The students thought that the worms would develop and reproduce better in orange waste, after papaya and guava (Fig. 20). However, when they followed the practical activity and the resulting experiments, they saw that the results were different from what they had expected.

During the practical vermicomposting activity, and the subsequent monitoring of the experiments by the students, it was observed that, among the waste used, the ones that obtained the best yield were guava, papaya and orange, respectively (Fig. 21).

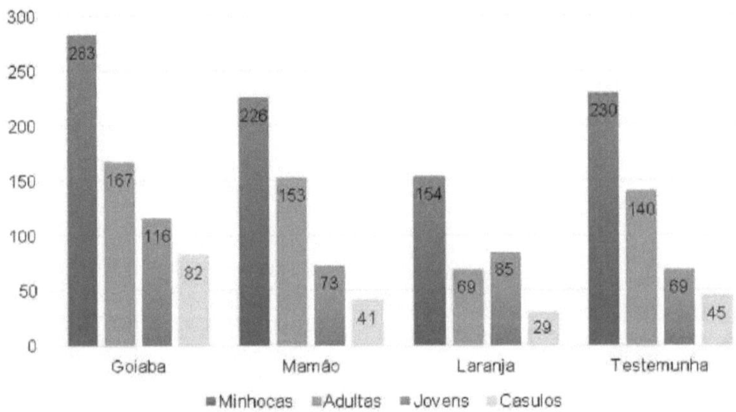

Figure 21 - Data from vermicomposting experiments resulting from practical lessons.

The difference between the results expected by the students and those that were actually confirmed when the practical activity was carried out, corroborates Bergamo (2010) and the students themselves, in understanding the importance of linking knowledge, of bringing theory and practice closer together.

3.2.3 Students' Conceptions after the Practical Activity

From expectations to the practical activity of vermicomposting, with subsequent monitoring of the experiments, there were significant changes in the students' worldviews. As we saw in the subsection above, what was expected did not materialize. The fruit preferred by the worms was not orange, but guava. The orange, as well as being preferred by them, proved to be the waste least used by the worms, and therefore the least effective for obtaining humus.

Even so, we can't deny the importance attributed to the dynamics operated by the students in the practical activity. If, in the theoretical classes, the students showed dispersion, since they hadn't assimilated all the content they had seen in class, this dispersion was reduced with the practical activity. In other words, the reports written by the students after the practical activity showed that they had appropriated the information they had learned during the activity (Fig. 22).

Figure 22 - Word cloud expressing the students' perspective after the practical lesson on the biological role and reproduction of oligochaetes.

In the word cloud (Fig. 22) drawn up from the students' reports, words such as worms, we, she, more, humidity, guava, better stand out. This shows that, compared to the pre-conceptions presented in section 3.2.1, there was greater assimilation of knowledge by the students when they reviewed the theoretical content in a practical activity.

Similarly, their testimonies at the end of the documentary demonstrate the necessary union of theoretical and practical didactic methodologies. For student Pedro, it was good to have taken part in the practical activity, he liked the "work, because I learned a lot in theory and in practice, because it's good for me to learn more about earthworms". Likewise, Augusto liked it very much

teacher Mariana's project. She taught us about vermicomposting and invited us to study with her. This week we observed the reproduction of earthworms, where we found cocoons, adult earthworms and adolescent earthworms. We look there and count how many worms have reproduced.

The expression of each of the students can be seen in the images of the practical activity. As can be seen in figure 23, the students participated significantly in this activity, given their motivation to carry out the activities proposed during the practical lesson.

Figure 23 - Images of the practical activity carried out with the students: **A)** Students, Horto Municipal agricultural technician and student Mariana; **B)** Detail of the worm farm. **Source:** author's archives (2015).

CHAPTER 4 WHERE WE STARTED FROM, WHAT WE UNCOVERED

"The young adult student arrives in the classroom full of theories, explanations and hypotheses. Their family, the community where they live, their work and their religiosity have allowed them to build up a wealth of knowledge. It is up to the teacher to find out what this body of knowledge is, made up of pure experience and perception, in order to use it to invite their students to access other ways of thinking, explaining, doing and acting (COSTA and BARRETO, 2006)".

4.1 Final considerations of the dissertation

The work carried out over the last few years has allowed us to gather a lot of information about science teaching and the study of oligochaetes. Theoretical immersion, as well as fieldwork, has made it possible to understand the theoretical and practical processes involved, sparking elements that shape our discussions throughout this dissertation.

Trying to find answers to our research problem, which is "*how can we provide students with a way of learning about the biological role and reproduction of oligochaetes, based on practical tests?*", was not easy, as we looked for information that has extremely subjective connotations in the subjects' worldviews. The research focused on the methodologies used by the teachers in their pedagogical dynamics. It also focused on the students' assimilation of the content taught by the teachers through these same methodologies.

Although we understand that the hypotheses have not been established as rigid structures - which deduce confirmation and refutation - at this point it seems fundamental to be guided by this thread in order to proceed with the final considerations of the dissertation.

In this sense, it is worth pointing out that the first hypothesis we put forward, which states that science teaching is mainly based on theoretical teaching schemes that do little to improve the assimilation of learning by the students, has been confirmed to a certain extent.

According to the results, a significant proportion of the lessons given by the teachers who make up the empirical universe are based on the dynamics of texts from textbooks or scientific books. This is due to the amount of time spent on the lesson, the discussions of certain content, the physical space made available to conduct the lesson, among other things.

Thus, many teachers choose to use theoretical methodologies rather than practical ones. Although they use auxiliary methodologies to the theoretical tools, such as the use of videos and/or documentaries and even practical activities, they often do little to address the situations that arise in the daily lives of students who are seeking knowledge and information.

Although the first hypothesis was partly confirmed, the second hypothesis was supported by the discussion of the results. In other words, despite the fact that science teaching is largely based on theoretical lessons, there are a variety of teachers who seek alternatives to the rigidity of this approach. We saw this in our immersion in the field.

It is common at the Alaôr Tarouco Municipal Agricultural Primary School for students to hold practical classes after the theoretical teaching, especially for subjects related to the students' training in agriculture - such as the practical activity on vermicomposting processes.

Bearing in mind that the research involved not only teachers from this school, who already develop some of the theoretical content linked to practical activities, but also other teachers from the municipal and state schools in the municipality of Piratini/RS, it is feasible to emphasize the importance of their participation in the practical activity of vermicomposting.

For those teachers who rarely left their classrooms, justifying it by the "lack" of space, the practical activity proved to be feasible and interesting to work on in their schools too. According to the testimonies recorded in the documentary, after taking part in the practical activity, many teachers took the idea of vermicomposting back to their work environment, achieving satisfactory results with the students and the whole school community.

However, the practical activity of vermicomposting also proved to be effective in the assimilation of knowledge by the students. As they themselves answered in the questionnaires, described in the report and reported in the documentary, this activity made it possible to assimilate the theoretical knowledge seen in class. In other words, the information from the theory class was better learned through the activity.

practical visualization of the reproduction and biological role of oligochaetes.

In view of this, the teacher's initiative to take students into everyday situations of their experience, as was the case with the practical activity exemplified by Vermicomposting, does provide greater and better quality in the assimilation of knowledge by them.

4.2 Lessons learned

In view of the information discussed in this dissertation, I am emphatic in stating that the daily life of the classroom, added to the motivation to study science, both on the part of the student and the teacher, with involvement, that is, integrating theory with practice, fascinates me, and it is for this reason that I dedicate myself to research in education, as well as the incessant search for new didactic alternatives for the classroom. The moments of greatest satisfaction and recognition in my life have been, and still are, mediating knowledge, seeing the sparkle in each student's eye and the learning as a result - signs of which are found in the day-to-day development of my master's studies.

May this sparkle in my eye, this motivation to always seek the best for the education of our children, young people and adults, enable me to take on other challenges, beyond the teaching practices I have known up to now and, to a large extent, internalized. May I be able, with the experiences I have acquired so far, especially the knowledge I have learned in this work, to walk the path of education with firm footsteps, bringing and exchanging knowledge with all the subjects who cross my life's path.

4.3 Contributions and Future Work

It's imperative to bear in mind that the end of something doesn't mean that the journey is over. On the contrary, when one cycle ends, another begins, because the march continues... And this is the idea that guides this dissertation.

We don't want to end this dissertation with a defense of the material prepared for the board. We hope that in the near future, this work will help people to develop didactic strategies based on everyday life, with a view to forming minds that are aware of and active in ongoing development processes.

For this reason, even before delivering the written version of this material, we set out to take it, albeit in oral and visual form (presentation of the documentary), to other institutions, such as the Sacred Heart of Jesus School, in the municipality of Pedro Osório. At this school, teachers and students discussed the biological role and reproduction of oligochaetes, the growing production of waste and mechanisms to minimize its harmful effects on our environment. Participation at this school was so successful that, as can be seen in figure 24, the event was posted on social networks.

On April 6, 2016, Mariana Lucas Moreira, a master's student at the Federal Institute of Rio Grande do Sul's Campus Pelotas Visconde da Graça, Post-Graduate Program in Science and Technology in Education, gave us a beautiful talk about her dissertation, presenting her documentary: An alternative and dynamic way of teaching Oligoquetas (earthworms). At the time, he referred to Pope Francis, who said that we should teach the language of the forehead, the heart and the hands. We need to think, feel and transmit values... and help!!! It's not just basic sanitation, but sanitizing our hearts, taking care of our common home, our Habitat. Our school received a copy of the documentary from the Master's student, which will be useful for teachers and family members alike. Thank you Mariana for your presence!!!

Figure 24 - Posting on social media of the participation of master's student Mariana Lucas at the Sagrado Coraçâo de Jesus School, in the municipality of Pedro Osório/RS.

Another place where the study was presented was the São José School, located in the center of Pelotas/RS. There, as well as presenting the documentary, students from the first year of secondary school were invited to give a course on the biological role and reproduction of oligochaetes, so that once they had mastered the content, they could pass it on to students from the third year of elementary school at the same school.

We also intend to take discussions about the biological role and reproduction of oligochaetes, i.e. their contribution to ecosystems (transforming organic waste into odorless fertilizer), to other schools and institutions such as the municipal network of Piratini/RS and the Department of Environmental Quality (SQA) of the municipality of Pelotas/RS, bringing projects to schools that minimize the disposal of waste from school meals in order to reuse it in vegetable gardens for food production. The idea is, through environmental education work, to enable schools in these municipalities to develop vermicomposting projects, so that students can use the theoretical knowledge gained in the classroom through practical activities.

Despite these initiatives, we know that they are still limited to the Pelotas region. As a means of dissemination, we will use the public virtual environment of the web to produce our documentary, so that our experiences are widely available and disseminated. Furthermore, if there are any invitations to work together, there is a willingness to build them collectively.

REFERENCES

AFONSO *et al.* **Qualitative analysis of the digital portfolio in the pedagogical training of health preceptors: advantages of *Wordle*.** 2011. Available at: <http://www.abed.org.br/congresso2011/cd/233.pdf>. Accessed Jan. 2016.

ANTUNES, C. **Professores e profauros**. Petrópolis: Vozes, 2013.

POINTER. **How to get to Escola Municipal Agropecuaria de Ensino Fundamental Alaor Tarouco - Centro**. Available at: <http://www.apontador.com.br/local/rs/piratini/escolas_publicas/XXQ2G4VW/escola_municipal_agropecuaria_de_ensino_fundamental_alaor_tarouco/como-chegar.html>. Accessed Jan. 2016.

ARANHA, M. L. de. **História da educaçâo**. 2 ed. Sâo Paulo: Moderna, 1996.

ASTOLFI, J.P. & DEVELAY, M.A. **Didàtica das Ciências**. Campinas, Sâo Paulo: Papirus, p. 121-130, 1990.

BAKKER, A.P. **Effect of earthworm humus and inoculation of the arbuscular mycorrhizal fungus Glomus macrocarpum Tul. & Tul. on the development of early-rooted cashew seedlings (Anacardium occidentale L.)**. Dissertation (Master's Degree) - Federal University of Ceará, Fortaleza, Brazil, 60p, 1994.

BARBOSA-LIMA, M. C.; CASTRO, G. F.; ARAÙJO, R. M. X. Ensinar, Formar, Educar e Instruir: A Linguagem da Crise Escolar. **Ciência e Educaçâo**, v. 12, n. 2, 2006, p. 235-245.

BERGAMO, M. The use of differentiated methodologies in the classroom: an experience in higher education. **Revista Eletrônica Interdisciplinar**, v. 02, n. 04, 2010, p. 01-10.

BOUCHÉ, M.B. **The establishment of earthworm communities. *In:* Satchell, J.E. (Ed.) - Earthworm ecology: from Darwin to vermiculture.**

London, 1983.

BROWN, G.G. and DOMiNGUEZ, J. **Use of earthworms as environmental bioindicators: principles and practices** - the 3rd Latin American Meeting on Ecology and Taxonomy of Oligoquetes (ELA- ETAO3). Acta Zoológica Mexicana, 26, 2010.

BROWN, G.G. and JAMES, S.W. **Ecology, biodiversity and biogeography of earthworms in Brazil.** *In:* **Brown, G.G. and Fragoso, C. (Ed.) - Earthworms in Latin America: Biodiversity and ecology**. Londrina, Embrapa Soja, 2007.

CABRERA, R.A.D. **Production of citrus seedlings in nurseries: Use of alternative substrate and inoculation with** *Xylella fastidiosa*. Dissertation (Master's Degree) - Luiz de Queiroz College of Agriculture, Piracicaba, 2004.

CARVALHO, A.M.P. & GIL-PÉREZ, D. **Formação de professores de ciências: tendências e inovações**. Sâo Paulo: Cortez, 120p, 2000.

CARVALHO, A.M.P. de. **Teaching Science: Uniting Research and Practice**. Sâo Paulo: Vozes, 2004.

CHASSOT, A.; OLIVEIRA J.R. **Ciências, ética e cultura na Educaçâo - Teaching Science in Brazil: A brief historical review**. Sâo Leopoldo: Ed. UNISINOS, 1998.

CLEMENTE, E. Presentation Studies in Antiquity. In: BAKOS, M. M.; CASTRO, I. B.; PIRES, L. A. (Org.). **Origens do ensino**. Porto Alegre: EDIPUCRS, 2000, 225f.

COMPAGNONI, L; PUTZOLU, G. **Cria moderna de lâs lombrices y utilización rentable del humus**. Barcelona: Editorial de Vecchi-S.A, 127p, 1985.

CORDEIRO, J. **Didâtica**. Sâo Paulo: Contexto, 186p, 2013.

COSTA, E.; BARRETO, V. **A sala de aula como espaço de vivência e**

Aprendizagem. Brasilia: MEC, 2006, 53p.

CRUZ, K. H. **A Educaçâo em seu Contexto Histórico: Desafios da Educaçâo Pùblica Brasileira Frente ao Terceiro Milênio**. Available at: <http://www.webartigos.com/artigos/a-educacao-em-seu-contexto- historico-desafios-da-educacao-publica-brasileira-frente-ao-terceiro-milenio/16486/#ixzz4EO94HAJ6>. Accessed Jan. 2016.

DAVIDSON, D.A. and GRIEVE, I.C. **Relationships between biodiversity and soil structure and function: evidence from laboratory and field experiments**. *Applied Soil Ecology,* 176-185, 2006.

DE-NARDIN, M. H.; SORDI, R. O. Attention learning: an opening to invention. **Revista Iberoamericana de Educación**, v. 04, n. 44, 2008, p. 02-11.

DEWEY, J. **Democracy and Education: an introduction to the philosophy of education, by John Dewey**. New York, NY: MacMillan, 1916.

EMBRAPA, Agrobiology. **Breeding earthworms to produce organic fertilizer capable of improving the soil's chemical, physical and biological attributes**. Available at:

<https://www.embrapa.br/documents/1355054/1527012/4b+-+folder+Minhocultura+ou+vermicompostagem.pdf/323fbedc-7b3c-4d89-bccd-70b490b8e88b>. Accessed on: Mar. 2016.

ESTEVES, J.V. **Substrate for vegetable seedling production with carbonized rice husk, organic compost and steam disinfection**. Horticultura Brasileira, Brasilia, v. 18, p. 805-806, 2000.

FERRUZZI, C. **Manual de lombricultura**. Madrid: Mundi-Prensa, 137p, 1996.

FOUREZ, G. **Crisis in Science Teaching? Investigaçôes em Ensino de Ciências**. v.2, p. 109-123, 2003.

FREINET, Célestin, **As Técnicas Freinet da Escola Moderna**. Lisbon: Editorial Estampa Ltda., 1975.

GIROUX, H. **Critical theory and resistance in education: beyond theories of reproduction**. Petrópolis: Vozes, 1986.

GRAZIANO DA SILVA, J. **A modernização dolorosa: estrutura agrària, fronteira agricola e trabalhadores rurais no Brasil**. Rio de Janeiro: Zahar, 1982.

HERNANDEZ, R.R. **La lombricultura y sus fundamentos**. Madrid, 302p, 1991.

HUBER, A.C.K. **Study of mesofauna (mites and springtails) and macrofauna (earthworms) in the vermicomposting process**.

Dissertation (Master's Degree) - FAEM/UFPel, Pelotas, 77p, 2003.

IBGE Cities. **Municipality of Piratini**. Available at: <http://cidades.ibge.gov.br/xtras/perfil.php?lang=&codmun=431460&search=rio-grande-do-sul|piratini>. Accessed in March 2016.

INEP. National Institute for Educational Studies and Research Anisio Teixeira. **Exploratory study on Brazilian teachers based on the results of the 2007 Basic Education School Census**. Brasilia: Inep, 2009, 63 p.

JAMES, S. **Earthworms (Annelida: Oligochaeta) of the Columbia River basin assessment area**. General Technical Report PNW-GTR- 491. Portland, Oregon, United States Department of Agriculture, 13p 2000.

JUSBRASIL. **Influence of Roman Law in the Present**. Available at: <http://ges.jusbrasil.com.br/artigos/153252615/influencia-do-direito- romano-na-atualidade>. Accessed in Jan. 2016.

KIEHL, E.J. **Organic fertilizers**. Sâo Paulo: Agronômica Ceres, 492p, 1985.

KNAPPER, C.F.U. **Associaçâo Brasileira de Minhocultura. ABRAMI**. Newsletter No. 3. São Leopoldo/RS, 1987.

LEE, K.E. **Earthworms: Their ecology and relationships with soils and land use**. Sydney, Academic Press, 411p, 1985.

LOURENÇO, N. **Caracteristicas da minhoca Epigea *Eisenia foetida* - Beneficios, caracteristicas e mais-valias ambientais decorrentes da sua utilização.** FUTURAMB Magazine, 1ª Ed. 2010.

LOURENÇO, N. **Manual de Vermicompostagem e Vermicultura para a Agricultura Orgânica**. Engebook, 1st Ed. 2013.

LUNETTA, V.N. **Actividades prâticas no ensino da Ciência**. Revista Portuguesa de Educaçâo, v.2, p. 81-90, 1991.

MADIGAN, M.T.; MARTINKO, J.M.; DUNLAP, P.V. and CLARK, D.P. **Brock's Microbiology**. 12th ed. Porto Alegre: Editora Artmed, 1160 p, 2010.

MARQUES, A. S. **The importance of approaches to earthworms in Teaching Science**. 2009. Available at: <http://adelsmarques.blogspot.com.br/2010/04/para-que-serve-as-minhocas.html>. Accessed Jan. 2016.

MARTINEZ, A.A. **Manual pràtico do minhocultor**. Jaboticabal: FUNEP, 137p, 1995.

MENDES, M. C. M. **The profile of the 21st century teacher: challenges and competencies**. 513f. 2011. Thesis (Doctorate in Educational Sciences) - Faculty of Educational Sciences, University of Granada, Granada, 2011.

MENEZES, J. G. de C. et al. **Estrutura e funcionamento da** Educaçâo **Bàsica leituras**. 2. ed. Sâo Paulo: Pioneira Thomson Learning, 2002.

MILONARI, R. B. **Vermicomposting and sustainability**. Available at: <http://www.ebah.com.br/content/ABAAABI6oAG/vermicompostagem-sustainability>. Accessed on: Jan. 2014.

MINAYO, M. C. **Pesquisa social: teoria, método e criatividade**. 24a ed. Petrópolis, RJ: Vozes, 1994. 80 p.

MORAN, José Manuel et al. **Novas tecnologias e mediação pedagógica**. 6. ed. Campinas: Papirus, 2000.

MORSELLI, T.B.G.A. **Practical Course in Vermicomposting**. Poligrafo. Bagé, RS, 1994.

MORSELLI, T.B.G.A. **Introduction to the study of organic waste. Handout for the subject of Organic Waste in Agricultural Systems**. Pelotas: UFPel/FAEM/PPGA/DS, 115p, 2005.

NEVES, E. Z. V. **From theory to practice: a bridge to be built from initial training**. 2012, 84F. Monograph (Degree in Languages - Spanish) - Federal University of Paraiba, Center for Human Sciences, Languages and Arts. Joâo Pessoa, 2012.

NÓVOA, A. **Concepçoes e prâticas de formaçâo continua de professores. In Formação continua de professores: realidades e perspectivas**. Aveiro, Portugal. University of Aveiro, 1991.

NUNES, M.U.C. **Production of vegetable seedlings using plasticulture and coconut**. Aracajù: Embrapa Tabuleiros Costeiros, 29p, 2000.

PAGANINI, E. L. Overcoming (in)security at the start of a teaching career. In: Seminars on Education Research in the Southern Region, 2012. **Electronic** proceedings of the IX ANPED Sul, 2012, p. 1-13.

PEREIRA NETO, J. T. **Composting Manual**. Belo Horizonte: UNICEF, 1996, 56p.

MUNICIPAL EDUCATION PLAN - PME. Available at: <http://www.prefeiturapiratini.rs.gov.br/sistema/uploads/documentos/documento_base_do_plano_municipal_de_educacao_-_piratini.pdf>.

Accessed Dec. 2015.

PRODANOV, C. C. **Metodologia do trabalho cientifico [recurso eletrônico]: métodos e técnicas da pesquisa e do trabalho acadêmico**. 2. ed. Novo Hamburgo: Feevale, 2013, 276p.

REINECKE, A.J., VILJOEN, S.A. **The influence of feeding patterns on**

growth and reproduction of the vermicomposting earthworm Eisenia foetida (oligochaeta). Bio. Fert. Soils, Oxford, v.10, p.184187, 1990.

RIBEIRO, M. Cooperative work in the MST and rural primary education: challenges for basic education. **Revista Brasileira de Educaçâo**, no. 17, 2001, p. 20-39.

RICCI, Marta dos Santos Freire. **Vermicomposting Manual**. Porto Velho: EMBRAPA, 1996.

RIGHI, G. **Oligochaeta (Annelida): diversity and agro-ecology**. *In* : Joly, C.A.; Bicudo C.E.M. (Ed.) - Biodiversidade do Estado de Sào Paulo, Brasil: sintese do conhecimento ao final do século XX. Sào Paulo, FAPESP, 1999.

RODRIGUES, V.C., THEODORO, C.A., ANDRADE, I.F., NETO, A.I.,

RODRIGUES, V.N., ALVES, F.V. **Earthworm production and mineral composition of vermicompost and feces from buffalo and cattle**. Ciência Agrotécnica. Lavras. V.27, n.6, p.14091418, nov./dez, 2003.

ROLA, M. O. R.; SILVA, R. F. Advantages of vermicomposting over traditional composting. **Revista F@pciência**, v.10, n. 1, 2014, p. 4048.

RUPPERT, E.E.; Fox, R.S. and BARNES, R.D. **Zoology of Invertebrates: a functional-evolutionary approach**. 7ª ed. Sâo Paulo, Roca, 1145p, 2005.

SCARLATO, F. C.; PONTIN, J. A. **Do nicho ao lixo, ambiente, sociedade e educaçao**. 17.ed. Atual Editora, 1993.

SELL, J C. **Different models, different paths: the search for environmental sustainability in the municipality of Piratini, RS**. 2011, 173f. Dissertation (Master's in Geography), Post-Graduation in Geography and Geosciences, area of concentration in Landscape, Environment and Environmental Quality, Federal University of Santa Maria (UFSM/RS), Santa Maria, 2011.

SOUZA, E.C.; SOUZA, I. C.; TEIXEIRA, V. R. **Historical evolution of the teaching-learning process** process.2014. Available at:

<http://www.seduc.mt.gov.br/Paginas/Evolu%C3%A7%C3%A3o-hist%C3%B3rica-do-processo-ensino-aprendizagem.aspx >. Accessed Jan. 2016.

SOUZA *et al.* The methodologies used by Science and Biology teachers in the teaching/learning process. **Revista da SBEnBio**, n. 07, 2014, p. 2014-2022.

TAGLIARI, P.S. **Agroecological production: an excellent alternative for family farming**. Agropecuària Catarinense, 10: 29-39, 1997.

TAVARES, R. **Significant Learning**. Revista Conceitos n.55, Sâo Paulo, 2004.

TEÓFILO, T. J. S.; DIAS, M. S. A. Teachers' and students' conceptions of teaching-learning methodologies: an analysis of the case of the Nursing Course at Vale do Acaraù State University in Sobral - Cearâ. **Revista Interface**, v.13, n.30, 2009, p. 137-51.

TIBAU, A.O. **Matèria orgànica e fertilidade do solo**. Sâo Paulo: Nobel, 172p, 1984.

WIKIPEDIA. **Piratini**. Available at: <https://pt.wikipedia.org/wiki/Piratini>. Accessed Jan. 2016.

MIX
Papier aus verantwortungsvollen Quellen
Paper from responsible sources
FSC® C105338

Printed by Books on Demand GmbH, Norderstedt / Germany